Review of the
Edwards Aquifer Habitat Conservation Plan

REPORT 2

Committee to Review the Edwards Aquifer Habitat Conservation Plan

Water Science and Technology Board

Division on Earth and Life Studies

A Report of
The National Academies of
SCIENCES • ENGINEERING • MEDICINE

THE NATIONAL ACADEMIES PRESS
Washington, DC
www.nap.edu

THE NATIONAL ACADEMIES PRESS 500 Fifth Street, NW Washington, DC 20001

This activity was supported by the Edwards Aquifer Authority under Contract No. 13-666-HCP. Any opinions, findings, conclusions, or recommendations expressed in this publication do not necessarily reflect the views of any organization or agency that provided support for the project.

International Standard Book Number-13: 978-0-309-45155-0
International Standard Book Number-10: 0-309-45155-8
Digital Object Identifier: 10.17226/23685

Additional copies of this publication are available for sale from the National Academies Press, 500 Fifth Street, NW, Keck 360, Washington, DC 20001; (800) 624-6242 or (202) 334-3313; http://www.nap.edu.

Copyright 2017 by the National Academy of Sciences. All rights reserved.

Printed in the United States of America

Suggested citation: National Academies of Sciences, Engineering, and Medicine. 2017. *Review of the Edwards Aquifer Habitat Conservation Plan: Report 2.* Washington, DC: The National Academies Press. doi: 10.17226/23685.

The National Academies of
SCIENCES · ENGINEERING · MEDICINE

The **National Academy of Sciences** was established in 1863 by an Act of Congress, signed by President Lincoln, as a private, nongovernmental institution to advise the nation on issues related to science and technology. Members are elected by their peers for outstanding contributions to research. Dr. Marcia McNutt is president.

The **National Academy of Engineering** was established in 1964 under the charter of the National Academy of Sciences to bring the practices of engineering to advising the nation. Members are elected by their peers for extraordinary contributions to engineering. Dr. C. D. Mote, Jr., is president.

The **National Academy of Medicine** (formerly the Institute of Medicine) was established in 1970 under the charter of the National Academy of Sciences to advise the nation on medical and health issues. Members are elected by their peers for distinguished contributions to medicine and health. Dr. Victor J. Dzau is president.

The three Academies work together as the **National Academies of Sciences, Engineering, and Medicine** to provide independent, objective analysis and advice to the nation and conduct other activities to solve complex problems and inform public policy decisions. The National Academies also encourage education and research, recognize outstanding contributions to knowledge, and increase public understanding in matters of science, engineering, and medicine.

Learn more about the **National Academies of Sciences, Engineering, and Medicine** at **www.national-academies.org**.

The National Academies of
SCIENCES • ENGINEERING • MEDICINE

Reports document the evidence-based consensus of an authoring committee of experts. Reports typically include findings, conclusions, and recommendations based on information gathered by the committee and committee deliberations. Reports are peer reviewed and are approved by the National Academies of Sciences, Engineering, and Medicine.

Proceedings chronicle the presentations and discussions at a workshop, symposium, or other convening event. The statements and opinions contained in proceedings are those of the participants and have not been endorsed by other participants, the planning committee, or the National Academies of Sciences, Engineering, and Medicine.

For information about other products and activities of the National Academies, please visit nationalacademies.org/whatwedo.

COMMITTEE TO REVIEW THE EDWARDS AQUIFER HABITAT CONSERVATION PLAN PHASE 2

DANNY D. REIBLE, *Chair*, Texas Tech University, Lubbock
JONATHAN D. ARTHUR, Florida Department of Environmental Protection, Tallahassee
M. ERIC BENBOW, Michigan State University, East Lansing
ROBIN K. CRAIG, University of Utah, Salt Lake City
K. DAVID HAMBRIGHT, University of Oklahoma, Norman
LORA A. HARRIS, University of Maryland Center for Environmental Science, Solomons
TIMOTHY K. KRATZ, University of Wisconsin, Madison
ANDREW J. LONG, U.S. Geological Survey, Rapid City, South Dakota
JAYANTHA OBEYSEKERA, South Florida Water Management District, West Palm Beach
KENNETH A. ROSE, Louisiana State University, Baton Rouge
LAURA TORAN, Temple University, Philadelphia, Pennsylvania
GREG D. WOODSIDE, Orange County Water District, Fountain Valley, California

National Academies Staff

LAURA J. EHLERS, Study Director, Water Science and Technology Board
BRENDAN McGOVERN, Senior Program Assistant, Water Science and Technology Board

WATER SCIENCE AND TECHNOLOGY BOARD

GEORGE M. HORNBERGER, *Chair,* Vanderbilt University, Nashville, Tennessee
EDWARD J. BOUWER, Johns Hopkins University, Baltimore, Maryland
DAVID A. DZOMBAK, Carnegie Mellon University, Pittsburgh, Pennsylvania
M. SIOBHAN FENNESSY, Kenyon College, Gambier, Ohio
BEN GRUMBLES, Maryland Department of the Environment, Baltimore, Maryland
ARTURO A. KELLER, University of California, Santa Barbara
CATHERINE L. KLING, Iowa State University, Ames
LARRY LARSON, Association of State Floodplain Managers, Madison, Wisconsin
DINAH LOUDA, Veolia Institute, Paris, France
STEPHEN POLASKY, University of Minnesota, St. Paul
JAMES W. ZIGLAR, SR., Van Ness Feldman, Potomac, Maryland

Staff
ELIZABETH EIDE, Acting Director
LAURA J. EHLERS, Senior Staff Officer
STEPHANIE E. JOHNSON, Senior Staff Officer
ED DUNNE, Staff Officer
M. JEANNE AQUILINO, Financial/Administrative Associate
BRENDAN R. McGOVERN, Senior Project Assistant

Preface

The Edwards Aquifer in south-central Texas is an important drinking water resource but also provides critical habitat for threatened and endangered species that inhabit the freshwater that forms the San Marcos and Comal Rivers. The unique habitat afforded by these spring-fed rivers has led to the evolution of species that are found in no other locations on Earth. Because of the potential for variations in spring flow due to both human and natural causes, the Edwards Aquifer Authority (EAA) and stakeholders have developed a Habitat Conservation Plan (HCP) to protect these unique threatened and endangered species. The HCP seeks to effectively manage the river-aquifer system to ensure the viability of the endangered species in the face of future water quantity concerns, such as drought and increased demand from population growth, as well as water quality threats to the aquifer.

The National Academies of Sciences, Engineering, and Medicine were asked by the EAA to assist in this process by forming a committee of expert volunteers that could review the implementation of HCP activities. The National Academies' study was planned in three phases, with the first phase being a review of the scientific efforts conducted to help build a better understanding of the river-aquifer system and its relationship to the endangered species, including monitoring and modeling. The first phase led to a report published in 2015 that provided an evaluation and recommendations for strengthening those efforts.

The current report is the culmination of the second phase. This report reviews the progress in implementing the recommendations from the Committee's first report, seeking to clarify and provide additional support for implementation efforts where appropriate. The current report also reviews selected Applied Research projects and minimization and mitigation (M&M) measures to help ensure their effectiveness in meeting the goals of the HCP. This report does not evaluate the adequacy of the goals and objectives of the HCP to protect the endangered species nor the capability of the M&M measures to meet those goals. These topics are expected to be part of the third and final report.

A committee was established under the auspices of the Water Science and Technology Board (WSTB) of the National Academies with the title Committee to Review the Edwards Aquifer Habitat Conservation Program. The Committee included 12 individuals representing expertise in all areas relevant to the Statement of Task, including the hydrogeology of the aquifer and the physics, chemistry, and biology of river systems. Four meetings were held over the past year since the release of the Committee's first report. The first two meetings were held in San Antonio and included presentations on current EAA and HCP activities relevant to the Statement of Task. I would like to thank the following individuals for giving presentations to the Committee during one or more of its meetings: Nathan Pence, Executive Director of the Habitat Conservation Program, EAA; Alicia Reinmund-Martinez, EAA; Jared Morris, EAA; Jim Winterlee, EAA; Mark Hamilton, EAA; Ed Oborny, BIO-WEST; Tim Osting, AquaStrategies; Mark Enders, the City of New Braunfels; George Ward, University of Texas; Bill Grant and Rose Wang, University of Texas; Todd Swannack, Engineer Research and Development Center; and Thom Hardy, Texas State University. I would also like to thank the many people who helped organize and run the three field trips taken by the Committee, particularly Nathan Pence, EAA; Ed Oborny, BIO-WEST; Zac Martin, City of New Braunfels; Melani Howard, City of San Marcos; and Steve Bereyso, SAWS.

Although Committee members represented many diverse perspectives and areas of expertise, which varied from river-aquifer hydrology to biology, we reached consensus on all recommendations included in the report. We hope that the EAA will find these recommendations useful as they guide the scientific initiatives designed to provide a solid foundation for effective management of the river-aquifer system and protection of the endangered species.

This report has been reviewed in draft form by individuals chosen for their diverse perspectives and technical expertise. The purpose of this independent review is to provide candid and critical comments that will assist the institution in making its published report as sound as possible and to ensure that the report meets institutional standards for objectivity, evidence, and responsiveness to the study charge. The review comments and draft manuscript remain confidential to protect the integrity of the deliberative process. We wish to thank the following individuals for their review of this report:

James J. Anderson, University of Washington, Seattle
John D. Bredehoeft, *NAE*, The Hydrodynamics Group, LLC., Sausalito, CA
Stephen R. Carpenter, *NAS*, University of Wisconsin, Madison
Wendy D. Graham, University of Florida Water Institute, Gainesville
Jessie C. Jarvis, University of North Carolina, Wilmington
Keith P. Johnston, *NAE*, University of Texas, Austin
Stavros S. Papadopulos, *NAE*, S.S. Papadopulos & Associates, Inc. Bethesda, MD
Steven F. Railsback, Lang Railsback & Associates, Arcata, CA

Although the reviewers listed above have provided many constructive comments and suggestions, they were not asked to endorse the conclusions or recommendations nor did they see the final draft of the report before its release. The review of this report was overseen by Patrick L. Brezonik, University of Minnesota, Minneapolis; and R. Rhodes Trussell, Trussell Technologies, Inc., who were responsible for making certain that an independent examination of this report was carried out in accordance with institutional procedures and that all review comments were carefully considered. Responsibility for the final content of this report rests entirely with the authoring Committee and the institution.

> Danny D. Reible, *Chair*
> Committee to Review the Edwards Aquifer
> Habitat Conservation Plan

Contents

SUMMARY	1
1 INTRODUCTION	11
2 HYDROLOGIC MODELING	25
3 ECOLOGICAL MODELING	49
4 BIOLOGICAL AND WATER QUALITY MONITORING	73
5 APPLIED RESEARCH PROGRAM	81
6 MITIGATION AND MINIMIZATION MEASURES	101
ACRONYMS	123

APPENDIXES
A Evaluation of the Predictive Ecological Model for the   125
   Edwards Aquifer Habitat Conservation Plan: An Interim
   Report as Part of Phase 2
B Biographical Sketches of Committee Members and Staff   159

Summary

The Edwards Aquifer in south-central Texas is the primary source of drinking water for over 2.3 million people in San Antonio and its surrounding communities, and it supplies irrigation water to thousands of farmers and livestock operators in the region. A karst aquifer with extremely high-yield wells and springs and rapid groundwater transport, the Edwards responds quickly to both rainfall events (known as recharge) and withdrawals, such as pumping for irrigation and water supply. The two largest springs emanating from the Edwards Aquifer are home to a number of endemic fish, amphibians, insects, and plants found nowhere else in the world. Because of the potential for reduced spring flow during drought, which the region has suffered from periodically, eight of these species are listed as threatened or endangered under the federal Endangered Species Act: the fountain darter, the San Marcos gambusia (presumed extinct), the Texas blind salamander, the San Marcos salamander, the Comal Springs dryopid beetle, the Comal Springs riffle beetle, the Peck's Cave amphipod, and Texas wild rice.

To protect the listed species, the Edwards Aquifer Authority (EAA) and four other entities created a 15-year Habitat Conservation Plan (HCP), which outlines a broad array of programs that when implemented will help to maintain the endangered species while managing withdrawals from the aquifer. The programs that make up the HCP range from long-term biological monitoring of the springs to restoration of native aquatic vegetation to the building of mechanistic models of the aquifer region. Given the diversity and complexity of the HCP, in 2013 the EAA requested the input of the National Research Council (NRC) during plan implementation. This report is the second of a three-phase study to provide advice to the EAA on various scientific aspects of the HCP. The Committee convened to conduct the study addressed the following tasks (with the chapters containing the material indicated in parentheses):

- Evaluate progress and modifications implemented as a result of the Committee's first report. *(Chapter 2 Hydrologic Modeling, Chapter 3 Ecological Modeling, and Chapter 5 Applied Research Program)*
- Continue to assess the methods of and data collected through the water quality monitoring and biomonitoring programs. *(Chapter 4)*
- Identify those biological and hydrologic questions related to achieving compliance with the HCP's biological goals and objectives that the ecological and hydrologic models should be used to answer, specifically including which scenarios to run in the models. These questions shall help generate information needed to make the HCP Phase 2 strategic decisions about the effectiveness of minimization and mitigation measures. *(Chapter 2 Hydrologic Modeling, Chapter 3 Ecological Modeling, and Appendix A Ecological Model)*
- Provide an evaluation of how the Phase 1 minimization and mitigation measures in the HCP (including flow protection measures and habitat restoration measures) are being implemented and monitored. Specifically, the Committee will discuss if the proper method of implementation is being utilized to achieve the maximum benefit to the covered species. *(Chapter 6)*

The reader is referred to (1) Chapter 1 for a description of the hydrology and ecology of the Edwards Aquifer and its springs, events that led to the creation of the HCP, and the plan's many elements; and (2) subsequent chapters for conclusions and recommendations not found in this summary.

HYDROLOGIC MODELING

The HCP calls for improvements to existing groundwater models of the Edwards Aquifer so that they can predict the effects of future hydrologic conditions (such as climate change and droughts) on spring flow and predict how management actions (like conservation measures) will affect water levels and spring flows. The Committee's first report recommended devoting future resources to a single model that incorporates the best concepts from existing models, rather than developing two "competing" models. It suggested that whatever model is selected should have features that advance the conceptual model of the system, such as telescoping meshes to accommodate shorter time scales and features for representing conduits and barriers. The Committee also stressed the need to quantitatively assess and present model uncertainty in formal EAA documents, using one or more of the following techniques: conducting more explicit sensitivity analysis; validating the groundwater model by testing its predictive abilities using data from a time period not included in the model calibration; using additional calibration and validation metrics; using PEST predictive uncertainty analysis; using the ensemble method; and having confidence intervals presented with all modeling results.

Subsequent to the first Committee report, the EAA created a Five-Year plan for hydrologic modeling, the objective of which is the continued updating of the hydrologic model, including conducting uncertainty analysis with the ensemble method. The Five-Year plan involves further development of the MODFLOW model, but not the second model of the groundwater system based on FEFLOW. A goal for the EAA will be to incorporate the learnings from the FEFLOW effort while maintaining a focus on the MODFLOW model. In addition, the Committee hopes that more of the improvements recommended in its first report are incorporated into the modeling effort, including more emphasis on conceptual model improvements, more careful evaluation of recharge estimation, further extension of uncertainty analysis, and improved descriptions of the modeling plans. Finally, several scenarios are suggested for the hydrologic model to improve its reliability and predictive capability.

The groundwater model should be tested against the 2011 to 2015 period, which was not used in model calibration. This period, which includes both very dry and very wet years, offers a remarkable opportunity to validate the model and enhance confidence in the model for future applications. Testing the model using the 2011-2015 period is likely to reveal the limitations of the current model. In addition, it should provide information on relative effects of withdrawals and effectiveness of management measures that were implemented during this period. The hydrologic, climatic, and well withdrawal data and the information on management actions for 2011-2015 should be more accurate than those from prior years, allowing for a more reliable assessment of the model.

Several scenarios are suggested for the hydrologic model, including optimizing the bottom-up package of the four spring flow protection measures,[1] evaluating spatial variations in pumping, and predicting how significant growth and land-use change in the recharge area might affect spring flows. Testing a variety of scenarios will not only improve the confidence in the model itself but will also help develop strategic decisions

---

[1] These measures are the Voluntary Irrigation Suspension Program Option, the Regional Water Conservation Program, Stage V Critical Management Period, and Aquifer Storage and Recovery.

associated with adaptive management and revisions to minimization and mitigation measures.

The Five-Year plan for the hydrologic model should include formal versioning and a decision support system that will be useful in future phases of HCP. The model should be updated every five years, with each new version including a peer-reviewed report and permanent archive of the numerical model that is available to the public. A decision support system will help minimize the subjectivity of management decisions that require a rapid response and should be included in Phase 2 of the HCP.

ECOLOGICAL MODELING

One of the major efforts set forth by the HCP is the creation of predictive ecological models for the Comal and San Marcos Spring systems. The models are expected to be able to account for impacts to the ecosystems from both management measures and natural variations, including such things as groundwater withdrawal, recreation activities, and restoration actions. The initial efforts of the ecological modeling team have focused on modeling the population dynamics of the fountain darter and key submersed aquatic vegetation (SAV) species.

In its first report, the Committee discussed the basic design of the fountain darter model, including the decision to develop an individual-based model, and it opined on several precursors to the model, such as the habitat suitability analyses done for fountain darter, Texas wild rice, and the Comal Springs riffle beetle. A subsequent interim report of the Committee, published earlier this year (see Appendix A), reviewed the first complete report from the ecological modeling team on what is now expected to be the sole product—models that predict the abundance of SAV and fountain darter, each run separately and also run in a coupled mode.

Chapter 3 addresses the EAA's response to its first report, and it suggests scenarios for the fountain darter model to run, now that a calibrated version is available. The comments and suggestions for scenarios presume that the recommendations in the interim report have been sufficiently addressed. In general, the Committee feels that the ecological modeling efforts have made good progress and that scientifically sound frameworks and approaches for the SAV and fountain darter models are in place. Model development is an iterative process. It is hoped that the models will continue to reflect new knowledge and understanding (beyond the originally anticipated time frame) in order to fully reap their benefits.

The EAA has now provided a scientifically sound basis for the development of a generalized ecosystem-based conceptual model. The conceptual diagrams produced to date for the fountain darter and SAV ecological models will help to guide further development of whole-system conceptual models. This collection of conceptual models will provide a communication tool for the HCP, will aid in coordination of the diverse expertise found across EAA's multiple advisory committees and contractors, and will serve an important function, along with the predictive ecological models, to evaluate the appropriateness and efficacy of the minimization and mitigation measures.

Armed with a fully capable fountain darter model, the scenarios analyzed should be designed and documented consistent with several concepts. These include careful designing of the scenarios and use of terminology to ensure transparency, confirming scenarios are within the domain of applicability, associating uncertainty with model predictions, and properly interpreting predictions and providing model-based mechanistic explanations for model responses.

Seven scenarios are described for the fountain darter model, which can be either diagnostic based (e.g., varying process rates) or evaluative (e.g., running the bottom-up package). The scenarios offered demonstrate how the model can be used to examine how extreme flows, process rates, environmental factors, SAV habitat, and episodic population reductions affect fountain darter population dynamics. These results can then be merged with the expected effects of minimization and mitigation measures to identify the robustness and redundancies of the entire suite of actions.

Only general guidance is given on possible scenarios for the SAV model, as it is not appropriate to provide detailed advice at this stage of development. Nonetheless, given the recently proposed adaptive management actions related to changing SAV species coverage goals in the HCP, it would be timely to evaluate the longer-term impact of these decisions on the stability of the SAV populations. The prospect of having such a valuable quantitative tool to better understand the effects of minimization and mitigation measures and predict future states will hopefully motivate those involved to continue developing the SAV model.

BIOLOGICAL AND WATER QUALITY MONITORING

The biological and water quality monitoring programs are intended to provide the observational data needed to assess whether the HCP is meeting its goals of protecting the covered species. Monitoring in the Edwards Aquifer spring systems has been ongoing since 2000 and is now even more comprehensive as a result of the HCP.

In its first report, the Committee commented on the design, purpose, integration, and adequacy of the two monitoring programs. In particular it raised concerns about the lack of integration between the water quality and biological monitoring programs, the difficulty of making system-wide estimates of target species population densities and trends given the reliance on non-randomized sampling of selected index reaches, the inability to assess whether changes in nutrient status are leading to changes in the frequency and magnitude of algal blooms because of insufficient detection limits of phosphorous and nitrogen, and the inability to determine population densities of invertebrates such as the Comal Springs riffle beetle.

In response, the EAA established two working groups to assess the water quality and biological monitoring programs, respectively, and make necessary modifications. Many of the Committee's recommendations were addressed, and the Comal Springs riffle beetle was made the subject of all Applied Research for 2016. Although the Committee feels that the monitoring programs remain strong, it identified areas for continued improvement. The following recommendations should be considered under the overarching goal of integrating the water quality and biological monitoring programs into a single program that provides the basic information needed to assess compliance with the HCP.

The monitoring program should include the measurements needed to monitor the performance of the broad suite of minimization and mitigation measures. Relying on the individual Applied Research projects or minimization and mitigation activities to provide these data is unrealistic, as these projects and measures are not designed nor funded over the long term, even though it may well take multiple years for the effects of these projects to be realized.

The monitoring program should include the long-term data required to test and inform continuous refinements of the ecological model. The ecological model will need to be continuously assessed and refined, and long-term data collected by the monitoring program will be critical to this effort. It is important that the modeling team be involved in the design of the monitoring program to ensure that the variables being measured are the ones that are most important for model assessment.

The EAA is making progress on addressing the sampling deficiencies that may limit the ability to estimate the distribution and abundance of the Comal Springs riffle beetle populations. The focus on the beetle in the Applied Research Program is a substantial effort for gaining knowledge about the distribution and life history features that will be important for understanding how the beetle responds to environmental variation, including changes in flow and responses during drought conditions. If the Comal Springs riffle beetle is to remain an indicator taxon for other listed invertebrate and vertebrate species, these gaps in life history and distribution will need to be addressed. Alternatively, the EAA should begin to develop monitoring plans for the other listed species.

APPLIED RESEARCH PROGRAM

The Applied Research Program created by the HCP has several goals, including filling gaps in knowledge about particular listed species, increasing understanding of key processes that affect their population dynamics, and providing data and information that can be used to parameterize and validate the ecological models. The overall goal of the program is to generate useful information during Phase 1 of the HCP to be able to make well-informed decisions about the overall direction of the HCP during Phase 2. Projects to date have been evenly split between the fountain darter, Texas wild rice, other SAV species, and the Comal Springs riffle beetle.

In its first report, the Committee provided a number of broad recommendations and conclusions about the Applied Research Program covering three general areas: improving the process used to solicit, review, and manage the Applied Research Program; adopting and implementing a data management system; and increasing understanding of the Comal Springs riffle beetle. The Committee was pleased that the EAA responded positively to its recommendations in these areas and has continued to devote resources to this program. Starting in 2018, the Applied Research Program will be used as a mechanism to assess the effectiveness of minimization and mitigation measures such as removal of exotic species, SAV restoration, and sediment control. The following additional conclusions and recommendations are made for the Applied Research Program.

The Committee applauds the changes made by the EAA regarding the procedures to identify, solicit, and review the projects in the Applied Research Program. The program as modified should be continued and could be expanded to facilitate additional multi-year studies in the future. To encourage more involvement of outside experts, the EAA should look for ways to ease barriers to participation in the Applied Research Program.

The Committee is supportive of EAA's attempts to develop an effective database management system that will provide data storage, curation, and access into the future. Resources for ongoing data management activities will need to be allocated throughout the lifetime of the HCP.

Monitoring the effectiveness of minimization and mitigation measures such as removal of exotic species, sediment control, and riparian conservation should be done through integration into the existing biological and water quality monitoring programs, rather than through one-off studies conducted through the Applied Research Program.

Modeling efforts should become more integral to consideration of future Applied Research projects. Projects in the Applied Research program can provide data and information to help design model scenarios, to improve parameter estimation and model formulation, and to enable model calibration and validation. For example, the Committee's previous recommendations that nutrients be considered in the ecological submodel of SAV would be easier to implement with nutrient data collection and more explicit consideration of nutrients in Applied Research projects on SAV.

MINIMIZATION AND MITIGATION MEASURES

The HCP lists 38 minimization and mitigation measures that when implemented are meant to protect the listed species from the impacts of both anthropogenic and natural disturbances to the Edwards Aquifer spring systems. Chapter 6 reviews the following minimization and mitigation measures and their implementation to date:

- SAV restoration/invasive plant removal in both the Comal and San Marcos systems
- Sediment removal at specific locations
- Dissolved oxygen management in Landa Lake
- Voluntary Irrigation Suspension Program Option (VISPO)
- Regional Water Conservation Program (RWCP)
- Stage V Critical Management Period
- Aquifer Storage and Recovery (ASR)

The first three measures were specifically identified for review by the EAA as a result of uncertainties about their implementation. The latter four, spring flow protection measures, were selected because of their importance to reaching the biological goals and objectives of the HCP.

In general, the Committee feels that implementation of key minimization and mitigation measures is moving in the right direction, with the various programs being characterized by competent project teams, sustained effort, and adequate initial performance monitoring. **For every minimization and mitigation measure implemented, performance monitoring should be done not only for the first year, but regularly during implementation, with a comprehensive synthesis of the monitoring data about every five years that goes beyond the simple trends analyses found in the HCP annual reports.** The following recommendations pertain to individual minimization and mitigation measures. Details can be found in Chapter 6.

SAV Removal and Restoration. Substantial progress has been made removing non-native vegetation from both the Comal and San Marcos systems and replacing it with native SAV species. Nonetheless, despite this sustained effort, **there is not enough new habitat from native plantings to maintain populations of fountain darter to balance non-native SAV removal.** This should be verified by considering the carrying capacity of the various SAV species (both native and non-native) for fountain darter.

Sediment Management. In general, **sediment removal activities should be limited to areas where ongoing upland sources or natural stream dynamics will NOT lead to deposition of new sediment within a matter of years.**

Dissolved Oxygen Management in Landa Lake. **The Committee recommends that aeration not be used routinely as a mitigation measure.** If floating mats cover more than 25 percent of the surface of Landa Lake and dissolved oxygen concentrations decrease, then manual breaking up and removal of the floating mats should be considered as a mitigation measure. Monitoring of dissolved oxygen concentrations using the miniDOT oxygen sensors in selected areas of Landa Lake and Upper Spring Run should be incorporated into an integrated water quality and biological monitoring program.

Voluntary Irrigation Suspension Program Option. **When the HCP is reviewed for renewal, it may be appropriate to re-evaluate the time period that the VISPO trigger is based on using a decision support system.** Consideration should be given to redefining the trigger to use additional information, such as groundwater elevation from a longer time frame, precipitation and recharge data, and groundwater model projections of future conditions.

Aquifer Storage and Recovery. **The Committee recommends that the following activities related to aquifer storage and recovery be initiated: (1) at a minimum of annually, determine specific injection at each ASR well to assess if there are any long-term changes in ASR well performance, (2) design and implement water quality monitoring for arsenic and related constituents in monitoring wells during recharge and storage events, and (3) design and implement water quality monitoring in ASR wells during recovery events.**

All Spring Flow Protection Measures. The total expense to implement the HCP in 2015 was $16,397,097, with the spring flow protection measures accounting for 67 percent of the total. Due to the high expense of the spring flow protection measures and their importance to the HCP's success, **the Committee recommends that compliance of the parties participating in the spring flow protection measures be audited** so that there is assurance that parties are complying with the terms of the program and the program will operate as designed.

1

Introduction

The Edwards Aquifer in south-central Texas is one of the most productive karst aquifers in the nation. Covering an area about 180 miles long and from 5 to 40 miles wide (see Figure 1-1), it is the primary source of drinking water for over 2.3 million people in San Antonio and its surrounding communities. The aquifer also supplies irrigation water to thousands of farmers and livestock operators in the region, which can account for as much as 30 percent of the total annual water withdrawals from the aquifer system. The Edwards Aquifer has extremely high yield wells and springs, with large volumes of groundwater being transported through the system on the order of days. Thus, the aquifer responds quickly both to rainfall events and to withdrawals, such as pumping for irrigation and water supply. The region has suffered periodically from droughts (most recently 2010-2014) that can be severe enough to reduce or halt flow at the major spring outlets. Indeed during the "drought of record" in the 1950s, flows at Comal Springs ceased for four months. If such reductions in spring flow were to recur, the results could be catastrophic to the organisms living in the Edwards Aquifer and its springs.

The two largest springs emanating from the Edwards Aquifer—Comal Springs in New Braunfels and San Marcos Springs in San Marcos—are home to a number of endemic fish, amphibians, insects, and plants found nowhere else in the world. Because of the potential for reduced spring flow during drought, eight of these species are listed as threatened or endangered under the federal Endangered Species Act (ESA): the fountain darter, the San Marcos gambusia (presumed extinct), the Texas blind salamander,

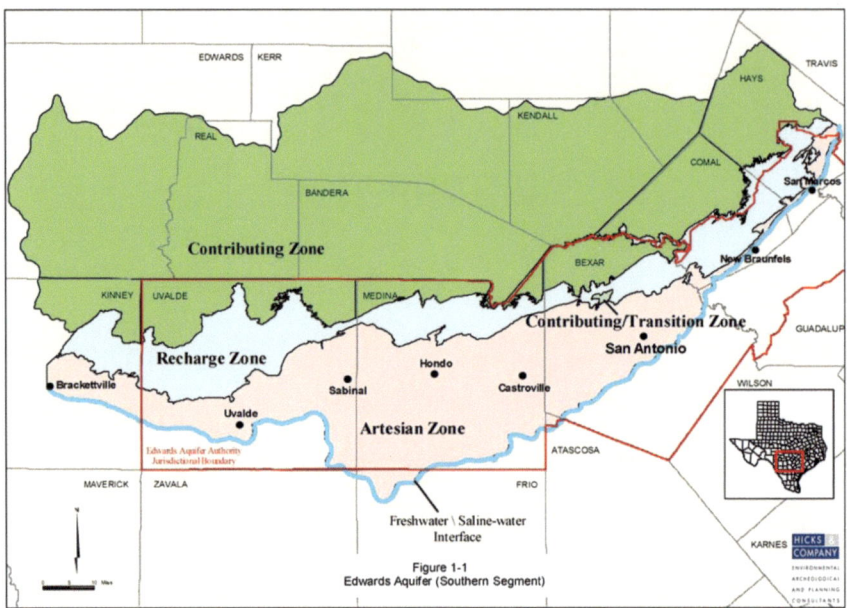

FIGURE 1-1 The Edwards Aquifer, showing the jurisdiction of the Edwards Aquifer Authority.
SOURCE: Figure 1-1 from EARIP (2012).

the San Marcos salamander, the Comal Springs dryopid beetle, the Comal Springs riffle beetle, the Peck's Cave amphipod, and Texas wild rice.

To protect the ESA-listed species, the Edwards Aquifer Authority (EAA) and four other local entities have created a 15-year Habitat Conservation Plan as part of their Incidental Take Permit under the ESA. The EAA is a regional government body tasked with managing domestic, industrial, and agricultural withdrawals from the Edwards Aquifer while maintaining spring flows at quantities that can support recreation and the ESA-listed species. The EAA implements the Habitat Conservation Plan, which the U.S. Fish and Wildlife Service (FWS) finalized and approved in 2013 after a years-long development process. Given the complexities of the Habitat Conservation Plan, in 2013 the EAA requested the input of the National Research Council (NRC) during implementation of the plan. This report is the second product of a three-phase study to provide advice to the EAA on various scientific aspects of the Habitat Conservation Plan that will ultimately lead to improved management of the aquifer. The first report (NRC, 2015) provides a comprehensive description of the hydrology and ecology of the Edwards Aquifer and its spring systems. It also describes in detail the events that led to the creation of the Habitat Conservation Plan and the plan's many elements. The reader is referred to Chapter 1 of that report for more in-depth information on these topics. A cursory summary is presented below.

THE EDWARDS AQUIFER

Hydrology and Climate

The Edwards Aquifer is a highly productive karst aquifer in south-central Texas. As shown in Figure 1-1, the contributing and recharge zones lie to the north, while pumping and artesian wells occur largely to the south. The largest area, the *contributing zone* (5,400 square miles), is where rainfall lands and is directed by streams toward the recharge zone. The *recharge zone* (approximately 1,250 square miles) is where precipitation percolates and flows into the groundwater to replenish the aquifer. In the *artesian zone* (2,650 square miles), the groundwater is under confined conditions, such that pressure levels in the aquifer cause the water to rise to elevations above the top of the aquifer. In such areas, groundwater flow from the aquifer to the land surface occurs in the form of springs and seeps. At least six springs occur within the artesian zone, including the two largest in Texas, the San Marcos and Comal Springs. Comal and San Marcos Springs are located within the San Antonio segment of the Edwards Aquifer, which spans approximately 3,600 square miles and is the focus of the Habitat Conservation Plan and this report. The karstic nature of the Edwards Aquifer, which is characterized by complex groundwater flow through such features as fractures, caves, and sinkholes, makes the aquifer vulnerable to potential surface water contamination.

The climate in the Edwards Aquifer region is characterized by significant spatial and temporal variability. Across the region, annual precipitation ranges from approximately 22 inches in the west to over 34 inches in the east. The mean annual precipitation for San Antonio from 1934 through 2013 was approximately 30.38 inches, although this varied annually by as much as 20 inches. Thus, it is not unusual for the Edwards Aquifer region to experience periods of high rainfall (in excess of 40 inches per year) separated by periods of drought. Evapotranspiration (unhindered vegetative rate) along the Edwards Aquifer region is similarly variable, ranging from more than 60 inches per year in the western extent to 30 inches per year in the eastern extent (Scanlon et al., 2005). However, recharge of the Edwards Aquifer occurs primarily via rapid, focused precipitation events entering the aquifer through exposed karst features within surface-exposed limestone, which lessens the role of evapotranspiration. Climate change scenarios suggest that, over the long term, precipitation in the region is expected to decrease and evapotranspiration is expected to increase (Loáiciga et al., 2000; Mace and Wade, 2008; Darby, 2010). Combined with an anticipated population increase and the associated increased demands on water resources, these factors suggest that the Edwards Aquifer is likely to become more stressed in the future.

Variations in climate in the Edwards Aquifer region are manifested in the variable nature of the aquifer's water budget. From 1934 to 2012, the median[1] annual recharge was 556,900 acre-ft,[2] with a range from 43,700 acre-ft during the drought of record in the 1950s to 2,486,000 acre-ft in 1992 (EAA, 2013). Edwards Aquifer discharge is composed of spring flows and consumptive use through wells. Total annual discharge from six of the most significant springs in the region monitored between 1934 and 2012 varied from 69,800 acre-ft in 1956 to 802,800 acre-ft in 1992, with a median annual discharge of 383,900 acre-ft (EAA, 2013). Well discharge estimates during the same period ranged from a low of 101,900 acre-ft in 1934 to a high of 542,400 acre-ft in 1989, with a median annual discharge of 327,800 acre-ft.

Ecology

Several species are endemic to the springs and river systems flowing from the Edwards Aquifer, including a variety of submersed aquatic vegetation (SAV), such as Texas wild rice; several fish, including the fountain darter; amphibians, such as the Texas blind salamander; and a variety of invertebrates. All species in the system depend on adequate spring flow, such that reduced flow in Comal and San Marcos Springs has periodically resulted in the intermittent loss of habitat and decreased populations. This loss of habitat from reduced flow is the main reason that eight species have been listed for protection under the federal Endangered Species Act (ESA) (see Table 1-1). Other threats to these species include increased competition and predation from invasive species, direct or indirect habitat destruction or modification by humans (e.g., recreational activities and reservoir construction), and other factors, such as high nutrient loading and bank erosion, that negatively affect water quality and habitat (USFWS, 1996).

---

[1] Note that the Committee recommends that means, not medians, be used in future reports on the water budget, including the Hydrologic Data reports from which this information was drawn.

[2] An acre-foot is the amount of water necessary to cover one acre of land with one foot of water. One acre-foot equals 1,233 cubic meters (m^3) of water.

TABLE 1-1 Common and Scientific Names of Species Proposed for Coverage Under the Edwards Aquifer Habitat Conservation Plan and Their Status According to the Endangered Species Act

Common Name	Scientific Name	ESA Status
Fountain Darter	*Etheostoma fonticola*	Endangered
Comal Springs Riffle Beetle	*Heterelmis comalensis*	Endangered
San Marcos Gambusia	*Gambusia georgei*	Endangered
Comal Springs Dryopid Beetle	*Stygoparnus comalensis*	Endangered
Peck's Cave Amphipod	*Stygobromus pecki*	Endangered
Texas Wild Rice	*Zizania texana*	Endangered
Texas Blind Salamander	*Eurycea rathbuni*	Endangered
San Marcos Salamander	*Eurycea nana*	Threatened
Edwards Aquifer Diving Beetle	*Haideoporus texanus*	Petitioned*
Comal Springs Salamander	*Eurycea* sp.	Petitioned**
Texas Troglobitic Water Slater	*Lirceolus smithii*	Petitioned

*Listed as under review by the USFWS
**Listed as undefined status by the USFWS

HABITAT CONSERVATION PLAN

The ESA, which in this case is enforced by the U.S. Fish & Wildlife Service (FWS), protects the listed species from actions that could jeopardize their continued survival. Most relevant to the EAA, the law prohibits the "take" of such species, which the ESA defines to mean "harass, harm, pursue, hunt, shoot, wound, kill, trap, capture, or collect, or to attempt to engage in any such conduct." The law also allows certain entities to apply for and receive an Incidental Take Permit, which defines the number of animals that can be "taken" by certain activities (such as groundwater pumping). In order for an applicant to receive such a permit, it must develop a Habitat Conservation Plan (HCP).

The HCP for the Edwards Aquifer took years to create and required the involvement of many parties (see NRC, 2015 for details). It was finally submitted by the EAA to the FWS in 2012, after which an Incidental Take Permit was issued. The permit will last 15 years, from March 18, 2013, until March 31, 2028. The five official Permittees are the EAA; the City of San Antonio, acting through the San Antonio Water System; the City of San Marcos; the City of New Braunfels; and Texas State University. All five have responsibilities under the HCP to implement minimization and mitigation measures that will protect the listed species and their habitat. The minimization and mitigation measures that make up the HCP include

(1) four spring flow protection measures, and (2) measures designed to maintain and restore the habitat of ESA-listed species at both Comal and San Marcos Springs. A complete list of the measures can be found in NRC (2015) or the HCP itself (EARIP, 2012). The discussion below focuses on the specific measures that are evaluated, in this report, for their ability to provide benefits to the listed species.

The four spring flow protection measures were designed to provide additional water during drought and include (1) critical period management, (2) regional water conservation, (3) a voluntary irrigation suspension program, and (4) aquifer storage and recovery. *Critical period management* refers to reductions in permitted discharges when the spring flow at Comal Springs and water levels at reference well J-17 fall below certain levels. To offset the risks to listed species under these conditions, the HCP instituted a new stage, Stage V, which would mandate reductions in pumping of 44 percent. The Regional Water Conservation Program builds upon the demand management already being conducted by the City of San Antonio. It is envisioned that new municipal conservation activities can save approximately 10,000 acre-ft/year (12.33 million m^3/year). The Voluntary Irrigation Suspension Program Option (VISPO) targets the 30 percent of annual Edwards Aquifer pumping that is withdrawn for irrigation. VISPO relies on permitted irrigators relinquishing their pumping rights when well levels and spring flows drop below certain triggers; it is intended to conserve another 40,000 acre-ft/yr (49.32 million m^3/year). Finally, the San Antonio Water System (SAWS) runs an aquifer storage and recovery (ASR) operation in the Carrizo Aquifer that will be expanded and is predicted to make the greatest contribution to overall Edwards Aquifer water savings (as much as 100,000 acre-ft/year or 123.3 million m^3/year).

Beyond spring flow protection measures, there are a variety of minimization and mitigation measures designed to maintain and restore the habitat of ESA-listed species at both Comal and San Marcos Springs. The measures that are evaluated in this report include aquatic vegetation restoration (including removal of invasive plant species and replanting of native species), sediment management in the spring and river systems, and dissolved oxygen management in Landa Lake.

Other programs found within the HCP that were the subject of the first report (NRC, 2015) and receive further attention in this report include water quality and biological monitoring of the aquifer and spring systems; improving the hydrologic model for the Edwards Aquifer; the creation of predictive ecological models for Comal and San Marcos Springs; and the Applied Research Program, which has been used to fund individual research projects to study the ecological dynamics within the Comal and San Marcos Spring systems. A brief overview of the first report is provided below.

THE EAA REQUESTED STUDY

In late 2013, the EAA formally requested the involvement of the NRC to provide advice on the many different scientific initiatives under way to support the HCP. An expert committee of the National Academies was asked to focus on the adequacy of the scientific information being used to, for example, (1) set biological goals and objectives, (2) determine what minimization and mitigation measures to use and their effectiveness; and (3) make decisions about the transition from Phase 1 to Phase 2 of the HCP. The study is being conducted from 2014 to 2018 and will produce three reports.

Phase 1 of National Academies Study

The Committee's first report (NRC, 2015) was released in late February 2015 and addressed four programs within the HCP: hydrologic modeling, ecological modeling, biological and water quality monitoring programs, and the Applied Research Program. In general, the report was complimentary of the efforts of the EAA and its partners in implementing the HCP and these four programs in particular, while at the same time identifying areas that could be improved upon.

Within the hydrologic modeling arena, which has been ongoing for decades, the Committee recommended devoting future resources to a single model that incorporates the best concepts from existing models, rather than developing two "competing" models. It suggested that whatever model is selected should have features that advance the conceptual model of the system, such as telescoping meshes to accommodate shorter time scales and features for representing conduits and barriers. The Committee also stressed the need to quantitatively assess and present model uncertainty in formal EAA documents. It mentioned several techniques for doing this, including conducting more explicit sensitivity analysis; validating the groundwater model by testing its predictive abilities using data from a time period not included in the model calibration; using additional calibration and validation metrics; using PEST predictive uncertainty analysis; using the ensemble method; and having confidence intervals presented with all modeling results.

Unlike the hydrologic modeling, the ecological model was new to the HCP; thus the Committee recommended creation of a conceptual model to help determine the most important processes for a model to encompass and to show the links between flow, species populations, and other important parameters. The initial species targets of the ecological model were fountain darter and SAV because of limited data and information about the other listed species. The Committee also stressed the importance of updating the habitat suitability analysis for Texas wild rice and of developing a much deeper understanding of the life history of the Comal Springs riffle beetle prior to including it in any ecological model.

With respect to the water quality and biological monitoring programs, the Committee was complimentary of the work that has been ongoing since 2000, which is now even more comprehensive as a result of the HCP. It noted that, because none of the sampling locations were selected using randomization procedures, results from the monitoring program are not representative of the entire spring and river systems and cannot provide system-wide estimates of population densities of target species. Enhanced sampling for nutrients and the development of new quantitative sampling methods for the Comal Springs riffle beetle were recommended.

The Committee suggested several new studies as well as programmatic issues that could improve the Applied Research Program. Examples of the latter include creating a more transparent process for prioritizing and funding Applied Research projects that includes stakeholder involvement and peer review; having greater competition and collaboration with outside scientific experts through open and widely disseminated solicitations for research; and offering some longer (e.g., two- to five-year) projects in order to maximize interest and collaboration from the region's leading researchers.

The Committee's first report closed with several overarching issues, including the need for more formal integration and database creation to enable clear explanation of the many sets of results emanating from the monitoring, modeling, and research efforts; the need to monitor the performance of minimization and mitigation measures currently being implemented; and the need for more formal and rigorous statistical analyses of laboratory and field data. Finally, it stressed the importance of considering various worst case scenarios as the HCP nears the end of its 15-year term.

After the release of NRC (2015), the EAA went through a lengthy process to determine how to implement the recommendations of the Committee. A formal document (EAA, 2015), including a prioritization matrix, was created by the newly formed Recommendations Review Work Group (RRWG). This document, the "Implementation Report," responded to every recommendation made by placing it into one of the following categories: (1) Done, (2) Continual, or (3) In Progress. The recommendations were also categorized as (4) To Be Implemented with No Budget Impact; (5) To Be Determined with Budget Impact; (6) To Be Determined If Implemented and Prioritized by Working Groups (Water Quality, Biological Monitoring, Applied Research); and (7) No, Not Recommended for Implementation. Not all categories are mutually exclusive. The responses of the RRWG are referred to periodically in all of the subsequent chapters when discussing how the EAA and others have responded to NRC (2015). While the Committee applauds the RRWG effort and the creation of the Implementation Report, it is not clear that current and future project plans and timelines will allow for the time and resources needed to address the recommendations in NRC (2015). The Committee recommends that project schedules be periodically revised to build in the time and resources needed to focus on implementing the recommendations from National Academies reports.

Phase 2 of National Academies Study

The statement of the task for the second phase of the National Academies review is given in Box 1-1. With respect to evaluating the progress and modifications implemented as a result of the Committee's first report (task no. 1), it is not our intent to exhaustively review all programs and list all recommendations given previously in NRC (2015). Rather, the relevant sections of Chapters 2, 3, 4, and 5 focus on specific issues where additional information or clarification may be helpful in improving the programs. The primary focus in this report is to help develop questions that both the hydrologic and ecological models can be used to answer, and to evaluate implementation of the minimization and mitigation measures that are being used to protect and restore habitat and protect flow. The question of

---

BOX 1-1
Review of the Edwards Aquifer Habitat Conservation Program—Phase 2
Statement of Task

A committee of the National Academies of Sciences, Engineering, and Medicine will:

1. Evaluate progress and modifications implemented as a result of the Committee's first report.
2. Continue to assess the methods of data collection and data collected through the water quality monitoring and biomonitoring programs.
3. Identify those biological and hydrological questions related to achieving compliance with the HCP's biological goals and objectives that the ecological and hydrologic models should be used to answer, specifically including which scenarios to run in the models. These questions shall help generate information needed to make the HCP Phase 2 strategic decisions about the effectiveness of conservation measures.
4. Provide an evaluation of how the Phase 1 conservation measures in the HCP (including flow protection measures and habitat restoration measures) are being implemented and monitored. Specifically, the committee will discuss if the proper method of implementation is being utilized to achieve the maximum benefit to the Covered Species.

whether these measures are sufficient and necessary to achieve the objective of ensuring survival of the listed species will be answered in the third and final National Academies report.

Hydrological Model

Tasks 1 and 3 in the Statement of Task (Box 1-1) pertain to the hydrologic modeling efforts that have been ongoing for some time in the Edwards Aquifer region. The current groundwater model for the Edwards Aquifer is a finite difference model based on MODFLOW that has been in development since 2000. (It should be noted that a finite element model of the aquifer was created during 2014 but it is not being developed further.) The major goal for the model is to be able to predict groundwater levels and spring flows under future hydrologic conditions, such as climate change and droughts. In particular, the model will be used to determine whether the four spring flow protection measures of the HCP can maintain flows at Comal and San Marcos Springs above levels critical to the listed species for sufficient durations.

Ecological Model

The ecological modeling described in the HCP has been under development since 2013 and is the subject of a short report from this Committee, released in June 2016 (NASEM, 2016; see Appendix A). According to the HCP, the two primary purposes for developing predictive ecological models are to identify and describe ecological responses of the listed species in the Comal and San Marcos Spring systems to various environmental factors and to predict and quantify impacts of various activities, including groundwater withdrawal, recreation, habitat restoration, etc., on these ecosystems and associated species. Mechanistic simulation models of moderate complexity were developed to simulate and predict the responses of fountain darter and SAV to changes in flow and water quality in the Comal and San Marcos Spring systems. The SAV model is still in the early stage of development, and the plan is for it to simulate the ecological processes of plant growth, mortality, and dispersal of multiple species on a spatial grid. The fountain darter model is further along in its development, with a functioning version now available. The fountain darter model is individual-based and uses the same spatial grid as the SAV model. Processes of growth, reproduction, movement, and mortality are represented for each individual fish. The spatial scale of both models is currently on the order of 1 m^2 with time steps ranging from hours or days. One goal for the future is to link the SAV and fountain darter models so that vegetation habitat in the fountain darter model also responds to flow and environmental variation, and ultimately

INTRODUCTION

to have the ecological models use the outputs of the groundwater model, so that simulations can be conducted for integrated surface and groundwater systems (EARIP, 2012). This report's treatment of the ecological modeling efforts directly responds to Tasks 1 and 3 in the Statement of Task.

Biological Monitoring

Task 2 in the Statement of Task specifically focuses on the biological and water quality monitoring programs of the HCP. A comprehensive biological monitoring plan was established by the EAA in 2000 to gather baseline and critical period data to fill important gaps in the ecological condition of the Comal and San Marcos spring and river ecosystems. This monitoring is ongoing during the 15-year term of the Incidental Take Permit in order to provide a means of monitoring changes in habitat availability and the population abundance of the listed species. The comprehensive monitoring plan increases in both frequency and the number of parameters examined as spring discharge falls below specific levels. The current program monitors the following components:

- Aquatic vegetation mapping, including Texas wild rice
- Fountain darter and fish community sampling
- San Marcos salamander sampling
- Comal Springs riffle beetle monitoring
- Comal Springs invertebrate sampling
- Comal Springs salamander sampling

A final goal of the biomonitoring program is to provide information to effectively determine whether the conservation measures are achieving the biological goals and objectives set forth in the HCP.

Water Quality Monitoring

Water quality monitoring has been in place in the Comal and San Marcos Spring systems for more than 40 years. The goals of the current program as it relates to the HCP are to detect water quality impairments that may negatively impact the listed species. Each year EAA monitors the quality of water in the Edwards Aquifer by sampling approximately 80 wells, eight surface water sites, and major spring groups across the region, including the Comal and San Marcos Springs. The program also includes sampling of stormwater runoff. Water samples are routinely analyzed in the field for selected water quality parameters (i.e., temperature, pH, conductivity, and alkalinity) and in the laboratory for common major ions, metals, total dissolved solids, hardness, bacteria, and nutrients. Many of the samples are also analyzed for semivolatile organic compounds and volatile organic compounds as well as pesticides, herbicides, and polychlorinated biphenyls.

Applied Research Program

As noted in Task 1 of the Statement of Task, the Committee continues to review the Applied Research Program, which was created by the HCP to (1) fill gaps in knowledge about particular listed species, (2) increase understanding of key processes that affect their population dynamics, and (3) provide data and information that can be used to parameterize and validate the ecological models. The overall goal of the program is to generate useful information during Phase 1 of the HCP to be able to make well-informed decisions about the overall direction of the HCP during Phase 2. Projects to date have been evenly split between species for which there is greater knowledge, like the fountain darter and Texas wild rice, and those for which less information is available, including SAV, the Comal Springs riffle beetle, and most of the other covered species. Critical to the recovery and protection of all aquifer species is knowledge of the species-specific demography and ecology, including knowledge of natural population fluctuations. Much of the research conducted under the Applied Research Program has been to better understand the ecological dynamics of the listed species under low flow conditions. The 2016 research projects are devoted exclusively to the Comal Springs riffle beetle.

Spring Flow Protection Measures

Task 4 in the Statement of Task brings a new topic under the Committee's purview, namely the ability of certain minimization and mitigation measures to provide benefits to the listed species. The four spring flow protection measures that are evaluated include the Voluntary Irrigation Suspension Program Option, the Regional Water Conservation Program, the Aquifer Storage and Recovery program of SAWS, and emergency withdrawal reductions during Stage V Critical Period Management. Each of these four measures is intended to contribute, in a cumulative fashion, to maintaining an adequate level of continuous spring flow during a repeat of the drought of record conditions (EARIP, 2012).

Habitat Restoration Measures

Beyond flow protection measures, there are as many as 29 other measures within the HCP aimed at restoring and improving the habitat of the listed species. Those that are considered in depth in this report fall into the following categories: native aquatic vegetation restoration, sediment management, and dissolved oxygen management. The measures designed to restore native vegetation include Texas wild rice enhancement and restoration in the San Marcos system (HCP section 5.3.1, 5.4.1); aquatic vegetation restoration and maintenance in both the Comal and San Marcos systems (non-native removal, native reestablishment) (5.2.2, 5.3.8, 5.4.3, 5.4.12); and management of floating vegetation mats and decaying vegetation and litter removal in both spring systems (5.2.4, 5.3.3, 5.4.3). Sediment management is focused to two areas in the San Marcos system: Sewell Park and Sessom Creek sand bar removal. Dissolved oxygen management is ongoing primarily in Landa Lake in the Comal system.

REPORT ROADMAP

Chapter 2 of this report addresses the hydrologic modeling of the Edwards Aquifer. It reviews how the EAA has responded to the recommendations in the first report regarding uncertainty analyses, recharge estimates, and how to represent features such as conduits and shorter time steps. It also poses questions that the hydrologic model could be used to answer as the HCP transitions from Phase 1 to Phase 2. Chapter 3 describes the ecological modeling for Comal and San Marcos Springs, focusing on the initial modeling efforts for the fountain darter and SAV. It builds on a short interim report (NASEM, 2016) released in June 2016 that dealt with the model structure and issues to keep in mind as the model is finalized (see Appendix A). The chapter finishes with scenarios that the ecological model should be used to address, which can either be diagnostic based (e.g., varying process rates) or evaluative (e.g., running the EAA's so-called bottom-up package of the four spring flow protection measures).

Chapter 4 updates the Committee's review of the comprehensive water quality monitoring program and biomonitoring program. It considers the adequacy of both programs and makes recommendations for what should continue to be sampled as the HCP moves forward. Chapter 5 discusses the Applied Research Program, including how the EAA has responded to recommendations in the first report. Chapter 6 comprehensively evaluates the four spring flow protection measures and the select habitat restoration measures of the HCP, considering how they should be implemented and monitored to be of maximum benefit to the listed species.

Each chapter ends with conclusions and recommendations that synthesize more technical and specific statements found within the body of each chapter. The most important conclusions and recommendations are repeated in the report summary. It should be noted that substantial information provided in the first report, such as the descriptions of each program, definitions of terms, and rationale for key recommendations, is not repeated in this report. The reader is referred to NRC (2015) for such details.

REFERENCES

Darby, E. B. 2010. The role of ESA in an atmosphere of climate change regulations. CLE International Conference: Endangered Species Act: Challenges, Tools, and Opportunities for Compliance, June 10-11, 2010. Austin, Texas.

EAA. 2013. Edwards Aquifer Authority Hydrologic Data Report for 2012. San Antonio, TX.

EAA. 2015. National Academy of Sciences—Review of the Edwards Aquifer Habitat Conservation Plan. Report 1 Implementation Plan. EAA August 20, 2015.

EARIP. 2012. Habitat Conservation Plan. Edwards Aquifer Recovery Implementation Program.

Loáiciga, H. A., D. R. Maidment, and J. B. Valdes. 2000. Climate change impacts in a regional karst aquifer, Texas, USA. Journal of Hydrology 227: 173-194.

Mace, R. E., and S. C. Wade. 2008. In hot water? How climate change may (or may not) affect the groundwater resources of Texas: Gulf Coast Association of Geological Societies Transactions 58: 655-668.

NASEM (National Academies of Sciences, Engineering, and Medicine). 2016. Evaluation of the Predictive Ecological Model for the Edwards Aquifer Habitat Conservation Plan: An Interim Report as Part of Phase 2. Washington, DC: The National Academies Press.

NRC (National Research Council). 2015. Review of the Edwards Aquifer Habitat Conservation Plan: Report 1. Washington, DC: The National Academies Press.

Scanlon, B., K. Keese, N. Bonal, N. Deeds, V. Kelley, and M. Litvak. 2005. Evapotranspiration estimates with emphasis on groundwater evapotranspiration in Texas. Prepared for the Texas Water Development Water Board, December, 2005, 54 pp. plus appendices.

USFWS. 1996. San Marcos and Comal Springs and associated aquatic ecosystems (revised) recovery plan. Department of Interior, U.S. Fish and Wildlife Service.

2

Hydrologic Modeling

EAA RESPONSE TO COMMITTEE'S FIRST REPORT

After the publication of the first National Research Council report (2015), the Edwards Aquifer Authority (EAA) established a Recommendations Review Work Group (RRWG) to identify and develop a plan for responding to all of the report's recommendations. The recommendations for the hydrologic model (see Chapter 1 page 17) included advancing the conceptual model of the system by using telescoping meshes to accommodate shorter time scales, better representing conduits and barriers, and quantitatively assessing and presenting model uncertainty in formal EAA documents. According to the RRWG (EAA, 2015a), some recommendations were already being implemented by the EAA, such as continued work on recharge estimation, others were mentioned as important to the EAA but work had yet to commence, while some recommendations are clearly not being worked on. In addition, the EAA created a Five-Year plan for hydrologic modeling. The objective of the Five-Year plan is the continued updating of the hydrologic model, including such steps as conducting uncertainty analysis with the ensemble method, documenting the model, and obtaining a peer review.

The Five-Year plan involves continued development of a single model, MODFLOW, although the Habitat Conservation Plan (HCP) mandated the creation of a second model of the groundwater system, which was based on FEFLOW. The RRWG did not resolve whether the EAA would move forward with one or both hydrologic models, but because the Five-Year plan prescribes use of the MODFLOW model only, it appears that the EAA's response to this recommendation has been resolved. The continuing challenge for the EAA is how to incorporate the learnings from the FEFLOW effort while maintaining a focus on the MODFLOW model.

According to the RRWG and the Five-Year plan, conceptual model changes are planned for future model versions, but not any time in the near future (such as the next five years). Nonetheless, the conceptual model changes suggested in NRC (2015) may substantially improve the quality of the model predictions and the model's usefulness as a planning and aquifer management tool, and some could be accomplished in the next five years. The section below suggests a path for evaluating and incorporating some of the most critical recommendations made in NRC (2015). The following are discussed: more emphasis on conceptual model improvements, further extension of uncertainty analysis, more careful evaluation of recharge estimation, and improved descriptions of the modeling plans. In general, the Five-Year plan would be greatly improved with additional detail, including better documentation of all activities involving the model and a timeline for specific model updates and improvements.

Improve Conceptual Model of Aquifer

Use Knowledge Gained from FEFLOW Effort

NRC (2015) recommended that improvements to the conceptual model of the Edwards Aquifer gained during the development of the FEFLOW model should be used in the future. The FEFLOW model was built using substantial resources, and it represents a considerable improvement to the physical representation of the system. In particular, the use of an unstructured grid, which permits the simulation of conduits explicitly, and the incorporation of the contributing zone into the model domain, are viewed as important enhancements. The calibration results of the FEFLOW model did reveal some inadequacies (Fratesi et al., 2015), and further work would be needed to improve the calibration. Thus, continued use of the MODFLOW model, which was further along in calibration, was determined to be the future course. Nonetheless, there are still opportunities to incorporate concepts from the FEFLOW effort into the current MODFLOW model. For example, extensive stratigraphic data that were compiled for the FEFLOW model could help inform knowledge of interformational flows, an uncertainty in the current model. In addition, the lessons learned from incorporating the contributing zone in FEFLOW will be useful for recharge estimation in the MODFLOW model and should be articulated now. Conduit and barrier features in the MODFLOW model were adjusted based on FEFLOW modeling, but additional evaluation of these features could be considered (see below). Some model runs, even if not rigorously calibrated, to help understand the sensitivity to these conceptual differences, would better support the model selection and prepare for the planned revisions in 2018 or 2019 to incorporate new features.

Test Conceptual Models Using Smaller Model Areas

In NRC (2015), the Committee made recommendations to both incorporate conduits and use telescoping grids. The latter recommendation provides a way to test conceptual models by taking advantage of parameter variation and calibration constrained to a smaller region. Thus, it is technically feasible to refine the model without excessive runtimes or cost, which was a concern of the RRWG. The refined model can be a gridded area nested within the larger model (a telescoping grid), or a separate model grid can be created. Indeed, the FEFLOW model examined subareas to focus calibration on smaller regions (Figure 2.4.2.3-2 from Fratesi et al., 2015).

Because telescoping grids are used to model a portion of the total area with a refined mesh, they make incorporation of conduits and finer time steps easier, as the model area under consideration is smaller. While there are additional costs associated with running these smaller models, they are small compared to the cost of the FEFLOW effort, which involved developing a second fully calibrated model. Modeling smaller areas can address some of the RRWG's concerns about cost and feasibility in testing conceptual models because there is no need to reconceptualize the entire HCP model.

Using telescoping grids or smaller model areas to explore model sensitivity to features such as conduits can help address systematic errors that have been observed in the current model. Predictions from the current model have a better match to observed values at low flows than at high flows. This type of systematic error tends to be caused by conceptual errors. A telescoping model could be used to compare different methods of incorporating conduits (e.g., high conductivity zones versus line elements in unstructured grids). A small region near key springs (e.g., using the estimated capture area for a given spring rather than the entire San Antonio pool) could be modeled with increasingly complex conduit patterns to see if they improve high flow calibration. Uncertainty about conduit locations does not need to limit the use of models to test concepts. For example, such models can be used to test whether simple or complex networks produce a match to spring flows and whether there are different networks activated at high flow in contrast to low flow. The sensitivity of the model to conduits and other heterogeneity patterns could thus be extended beyond what was proposed in the Five-Year plan (which does not clearly explain how conduits will be incorporated). In addition, finer time steps may provide improved prediction of both high and low discharge, but in particular improved prediction of high discharge in response to storms.

Because information is lost when coarse, simplified models are constructed, doing some modeling at a finer scale can inform the reliability of the large scale results. The refining of temporal and spatial scales is common practice in hydrologic modeling. NRC (2015) discussed the importance of improving conceptual model understanding by incorporating conduits into future modeling. It is likely that even extensive efforts to calibrate the model may not yield satisfactory results if the model's conceptual representation of the aquifer is not adequate. The use of sensitivity analysis on heterogeneity framed around conduit configurations and taking advantage of telescoping grids could help resolve long-standing debates about the role of conduits in model forecasts. These uncertainties need to be addressed to provide confidence in the models and bring the modeling up to current practices. Note that the type of conceptual model testing described here could be considered as modeling additional scenarios that help establish confidence in the model. The suggestions above do not fundamentally change the model or lead to development of a new model but simply allow the exploration of the model sensitivity to changes in the conceptual model of the aquifer.

Uncertainty Analysis

NRC (2015) describes five methods of uncertainty analysis that could be applied to a groundwater model of the Edwards Aquifer, which are showing error bars on spring-flow and water-level predictions, sensitivity analysis using the ensemble approach, testing the model's predictive abilities using data from a time period not included in the model (see subsequent section), PEST predictive uncertainty analysis (Brakefield et al., 2015), and data collection for reducing predictive uncertainty. These methods are listed in order from easiest to most difficult to implement.

The RRWG identified uncertainty analysis in the Five-Year plan, but only the ensemble approach is mentioned. That is, the Committee was presented with plans to use an ensemble approach in which about 10 variations of recharge estimates would be applied to the model (see details in Recharge section below). Recharge estimates are a good place to start an ensemble uncertainty analysis, because it is clear that recharge originating in the contributing zone and the interformational inflows are some of the most uncertain inputs and have a large influence on water levels and spring flows. However, using only one approach for uncertainty analysis does not line up with the report recommendation to explore more techniques and parameters.

In one of the first hydrologic modeling presentations to the Committee (Winterlee, 2014), the model results were shown for four recharge scenarios and results presented in terms of calibration targets. The current plan expands the recharge scenarios, but again presents results in terms of calibration targets with no statistical analysis (Winterlee, 2015). There was no indication that other conceptual-model parameters, boundary conditions, or other assumptions will be included in an ensemble approach for uncertainty analysis.

Another recommendation made in NRC (2015) was to display error bars in documentation of the groundwater model predictions; this was supported by the RRWG. Nonetheless, the Five-Year plan does not mention error bars, and modeling results shown at the committee meeting on February 2, 2016, did not incorporate them.

Although uncertainty analysis lends credence to models, arguments against its application include, among others, that it cannot be understood by policy makers and the public. However, Pappenberger and Beven (2006) explain why arguments such as this are not tenable. Uncertainty analysis can be used to illustrate possible ranges in the effects of system stressors and, therefore, enhances the interpretation of model results. Rather than letting the concern about public perceptions limit best practices in modeling, techniques should be applied to improve model design and data collection that decrease uncertainty (Anderson et al., 2015; Pappenberger and Beven, 2006). The potential for uncertainty analysis to increase transparency and lead to better decisions is exemplified by the case study of Enzenhoefer et al. (2014) summarized in Box 2-1.

Consideration of Recharge

NRC (2015) applauded the EAA's focus on refining estimates of recharge in the hydrologic modeling. At the Committee meeting held February 3, 2016, the EAA indicated their continued interest in exploring recharge by developing an ensemble of recharge estimates that incorporates variations on assumptions related to recharge mechanics. The specific details of the methods and assumptions that will be applied were not described to the Committee, other than to say that the spatial variability of recharge may vary between estimates. All the model parameters that were originally calibrated will be recalibrated for each of the ensemble simulations for recharge estimates. This is similar to the approach taken by Brakefield et al. (2015), a recent Edwards Aquifer model in which the need for recalibration is described in detail. The ensemble of recharge estimates will include modifications to estimates based on the Puente (1978) method (the use of which is required by the HCP). For example, the original Puente method resulted in peaks in simulated spring flow hydrographs that were much larger than observed values for some periods. Therefore, peaks in the recharge estimates will be arbitrarily reduced in the ensemble of recharge estimates.

Beyond Puente, there appear to be at least two additional recharge estimation methods that have been applied to the groundwater modeling

BOX 2-1
CASE FOR UNCERTAINTY ANALYSIS

MODFLOW and PEST were used to develop a wellhead protection model of a karst region in Germany (Enzenhoefer et al., 2014). Wellhead protection defines an area around a well that is vulnerable to contamination based on the capture area during pumping and is important for preserving water quality. Like the Edwards Aquifer, the study region had uncertainty in parameters due to karst, but a stakeholder group needed to make decisions about land use. Although this model was constructed for a smaller region than the Edwards Aquifer model, many of the concepts and tools are transferable.

Multiple realizations of variations in 13 calibration parameters (such as hydraulic conductivity for different zones and recharge) were modeled to produce a probabilistic result. These were presented as isoprobability contour maps (see Figure 2-1-1). These maps allow decision makers to see relative vulnerability of different areas and relate this to the costs of land-use restrictions. A map of probability of a travel time less than 50 days (a defined compliance level) showed that the 10% probability of exceedance extended beyond the existing protection zone in the west but there was more coverage than needed in the south. If only a 25% reliability is acceptable, then the existing protection area suffices. This method can also estimate the land area for a higher level of certainty along with associated cost, with an additional 1% reliability requiring only 0.08 km^2, but 6% additional reliability requires 0.78 km^2 at significantly more cost.

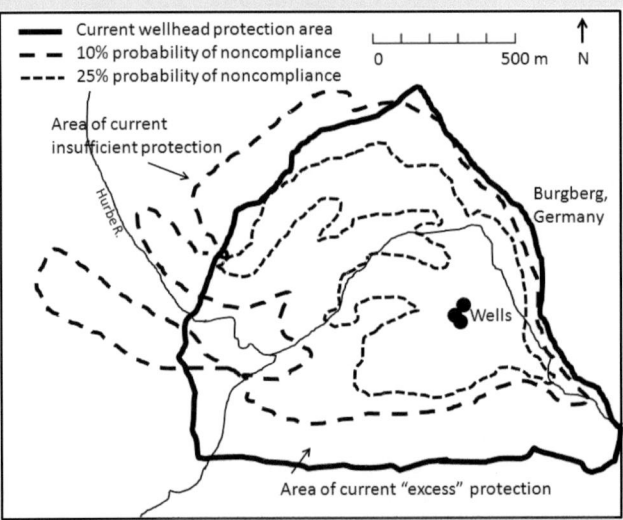

FIGURE 2-1-1 Probability map of 50-day travel time to well field, constructed using multiple realizations of 13 calibration parameters.
SOURCE: Adapted from Enzenhoefer et al. (2014).

efforts to date. First, a new recharge estimation method was applied to the FEFLOW model using NEXRAD[1]-estimated precipitation data. According to the FEFLOW final report (Fratesi et al., 2015), there were problems with using NEXRAD estimates, including that (1) the estimates were not available for 2001 to 2002 and (2) there were "data gaps and suspicious data." It is possible that these problems could have been avoided by using Daymet data (https://daymet.ornl.gov/ and http://cida.usgs.gov/gdp), which contains gridded weather parameters for the United States at a 1-km resolution for 1980 to the present. The data are based on weather-station data, and the spatial interpolation accounts for topography. At this time, it is not clear if this method will be further developed by the modeling team.

Second, the EAA spent considerable time developing recharge estimates using the Hydrologic Simulation Program—Fortran (HSPF) (http://water.usgs.gov/software/HSPF/). The prioritization matrix found in EAA (2015a) indicates that development of the HSPF model for recharge estimation will continue. However, no new progress on HSPF modeling since the first Committee meeting (February 2014) has been presented. The Committee recommends that recharge estimates from the HSPF method be included in the ensemble approach being used for uncertainty analysis (see section below).

Beyond these methods, there are other methods for estimating recharge that would enhance the ensemble, including a soil-water-balance (SWB) model developed by the U.S. Geological Survey (USGS) that estimates spatially distributed daily recharge on the basis of gridded weather and soils data (Westenbroek et al., 2010). Like the FEFLOW recharge estimation method discussed above, the SWB model utilizes Daymet data as input. The Committee recommends using as many different recharge estimation methods as feasible, and varying uncertain recharge parameters within these methods, to create the ensemble. The ensemble will provide a range of possible outcomes for spring flows; this range can be examined for calibration periods and validation periods, and, most importantly, for future scenarios predicted by the model.

Updating the MODFLOW Model: Adaptive Modeling

Although adaptive modeling was embraced by the RRWG, the next step in updating the conceptual model in the Five-Year modeling plan is not until 2019. That delay does not ensure that the hydrologic model uses the most recent tools and data available. The suggestions mentioned above would move the model further toward adaptive techniques. In addition, there are other steps that could be taken which include providing

---

[1] NEXRAD stands for Next-Generation Radar, a network of high-resolution weather radars operated by the National Weather Service.

more detailed plans, improving data management, and regular updating of the model with new field data. The Five-Year plan could provide more details about what updates are going to be incorporated. Providing more specifics about what updates will occur enhances communication.

The importance of collecting additional field data to improve the groundwater model was discussed in some detail in NRC (2015). For example, data collection can help to better understand the mechanics of flow toward springs, by characterizing conduits and evaluating hydraulic connections between the Trinity and the Edwards aquifers (which is a key part of the current groundwater research effort). Another example of data collection that would benefit the modeling is the incorporation of all available pumping data. Finally, variations in rainfall observed in the past few years would be enormously helpful in identifying strengths and weaknesses of the model; incorporating these data into model testing should receive a high priority. The Five-Year plan mentions assessing new data in 2017, but does not yet show an iterative approach between data collection and model updates. Field efforts are ongoing in other parts of the EAA and other organizations, such as the USGS and the Southwest Research Institute (SWRI). There should be a member of the modeling team who communicates regularly with the monitoring team about how current research can be incorporated into the model. Effective communication between data-collection staff and modeling staff is critical to maximize the effectiveness of both of these groups. Although long-term field research involves significant resources, such investment in bridging between field research and modeling will be cost effective in the long run.

While we urge including more details in the Five-Year plan for hydrologic modeling, we also recognize there is an inherent conflict between detailing a Five-Year plan and allowing for adaptive management of a model (i.e., updating as results and tools become available). The purpose of doing more planning is not to provide a road map that is immutable, but rather to provide a framework for discussing and improving the model. That is, it may be necessary to update the Five-Year plan more frequently than every five years (e.g., every two to three years) if new information becomes available and the original plan becomes outdated.

SCENARIOS FOR HYDROLOGIC MODELING

The MODFLOW model is expected to continue to be the primary groundwater modeling tool for the HCP. It is essential that the EAA strives to improve the predictive skills of the model for the anticipated refinements to the flow protection measures that may be necessary in Phase 2. Once the current improvements to the model are complete, it should be used to test a variety of scenarios, which will not only improve the confidence in the model itself but also will help develop strategic decisions associated with adaptive management and revisions to minimization and mitigation measures. The Committee's ecological modeling interim report (NASEM, 2016) discusses the importance of using models to test concepts and understand parameters and system conditions, not just produce predictions, which can be highly uncertain. Similarly, the hydrologic model is not just a tool to produce head maps and discharge values at springs that are compared to targets, but should also be used to evaluate scenarios that help understand what processes are important in the system.

The definition of the term "scenarios" in this section is broader than what is typically used for understanding system behavior under uncertainty. In view of the potentially significant economic costs associated with the four flow protection measures, it is extremely important to ensure that the model has adequate predictive skills under alternative future condition. Hence, the testing of the model against recent observations not used in the calibration (referred to as model validation) is suggested as a particular scenario to verify that the predictive skills of the model are acceptable and to determine if further improvements are needed. The first section below recommends testing the model against the most recent drought period (2011-2014) and the wet year of 2015. These years should have more accurate data (e.g., pumping) and management interventions that can enhance the confidence in the model as a predictive tool. Several modeling scenarios that can be run are then described, including lesser and more severe droughts as compared to the drought of record, optimization of the EAA's so-called "bottom-up package" of the four spring flow protection measures, understanding the influence of spatial patterns of pumping, and potential implications of land-use changes in the contributing zone on the water budget in the region. These scenarios are designed to ensure that the EAA meets the requirements of the adaptive management phase of the HCP by having a model that has been tested under a variety of conditions and by optimizing the flow protection measures in order to ensure that they meet specified goals.

Testing the Model by Comparing Predictions to the 2011-2015 Data

One of the recommendations for uncertainty analysis in NRC (2015) was to test the most recent version of the MODFLOW model against data from periods outside the period of calibration. The planned calibration period for the MODFLOW model is the period 2001-2011 (Winterlee, 2016). The model will then be used to evaluate the drought of record as a test of the accuracy of the model for simulating drought conditions. If the drought of record is not simulated adequately, the EAA may need to proceed with further calibration. If calibration is needed after attempting to validate the model, a description of what additional calibration was necessary should be documented in the model report. The ongoing efforts to improve the model, including a recalibration, should produce a better tool for future applications. This type of approach for using the model for conditions that have not been observed is not atypical and probably the only way to develop water resources management plans.

To improve the predictive skills of the model further, the MODFLOW model should be tested for periods outside the drought of record but under less extreme conditions and where more accurate data are available. The question that this scenario is trying to answer is "How accurately can the model predict conditions outside the time period used in the calibration?" In particular, the Commiteee recognizes the importance of simulating the more recent drought of 2011 to 2014 using current model parameters. These years are embedded in a longer-term drier period, which appears to have started in 2003 and which had a cumulative rainfall deficit of 82 inches (see Figure 2-1). The period is important as it includes more accurate well withdrawal data, better rainfall and flow information, extensive water level datasets, and, more importantly, the implementation of selected flow protection measures. For instance, the annual recharge during 2014 was only about 107,000 acre-feet, which was the second lowest recharge since 1934. On March 2012, the Uvalde pool reached the Stage V

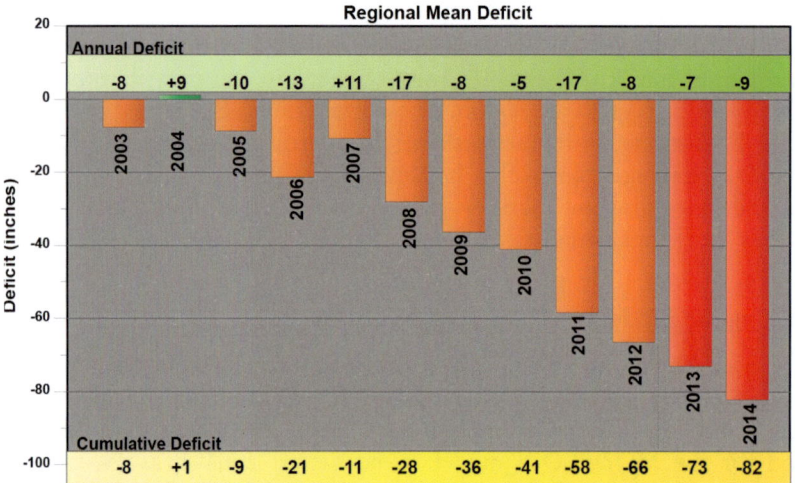

FIGURE 2-1 Regional mean deficit (annual and cumulative) rainfall from 2003 to 2014. The bars show the cumulative rainfall deficit for the period 2003-2014. SOURCE: EAA (2015b).

critical period management trigger for the first time. During this drought event, the precipitation was lower and groundwater usage was higher, yet spring flow did not dip quite as low in 2014 as it did during the 1950s. The difference may have been due to HCP conservation measures that were implemented during this period. Data and information gained during their implementation could be useful to test and improve the current model. It is important to note that the EAA may need to devote resources to acquire rainfall, pumpage, and other data necessary for simulating conditions during the 2011 to 2014 period. Finally, it would also be instructive to use the very wet year of 2015 to test model predictions (i.e., the recovery of the aquifer after a period of drought could be a sensitive test of model behavior).

The exercise of testing the model using the 2011 to 2015 period is likely to reveal the limitations of the current model. In addition, scenario testing should provide information on relative effects of withdrawals and effectiveness of management measures that were implemented during this period.

Performance of the System under a Variety of Drought Conditions

To date, the only scenario that the EAA has sought to run in the hydrologic model is the management program that includes the four spring flow protection measures (aka the bottom-up package) under conditions of the drought of record. These measures include the Voluntary Irrigation Suspension Program Option (VISPO), the Regional Water Conservation Program (RWCP), Aquifer Storage and Recovery (ASR), and Stage V Critical Management Period reductions. The model has simulated the effects of these measures on spring flow at Comal Springs during a repeat of the drought of record (Figure 2-7 in NRC 2015 and in many other documents). The model run used system demands that reflected the "permitted" withdrawals given by Initial Regular Permits. The total demand was approximately 572,000 acre-feet per year which was distributed spatially according to the 2008 pumping pattern in each county. The bottom-up package was also designed for the climatic conditions of the drought of record, the 1951-1956 period (HCP main report, and Appendix K). This period was the most severe drought recorded since 1934, and because of its duration and magnitude, the hydrologic effects were significant as was evident from the cessation of spring flow at Comal Springs for 144 days in 1956. Available evidence suggests that the six-year drought was indeed a very rare event, particularly with respect to duration. Droughts in the region are usually of shorter duration, although they could be more or less intense. From an extended record (280 years from 1700 to 1979) of Palmer Drought Severity Index (PDSI) developed using a correlation to tree ring data, droughts exceeding three years in duration occurred only four times, and three of them in the 1700s (EARIP, 2012). The drought of record was the fourth and most intense (EARIP, 2012).

Although the drought of record is a rare event (both in intensity and duration), the Committee had suggested that the 1950s drought may not represent the true worst-case scenario as the baseline for hydrologic modeling (NRC, 2015). Tree-ring data and other studies have indicated the possibility of more severe "mega-droughts." In view of the nature of the study and level of protection needed for the species that are threatened or endangered, it may be prudent to design a hydrologic scenario that simulates climatic and socioeconomic conditions which are more severe than what was used for the HCP. The question such a scenario is designed to answer is "How sensitive is the model to extreme conditions?" Another compelling reason for such an analysis is the vulnerability of the region to climate change. Mace and Wade (2008) and Loáiciga et al. (1996) have suggested that the Edwards Aquifer is the groundwater resource in Texas most vulnerable to climate change. While recognizing the lack of data for more severe drought scenarios, the use of paleo data (e.g., tree rings) and possibly stochastic modeling of rainfall patterns should be explored for the development of extreme scenarios. The climate scenarios should be designed considering the results of climate-model predictions available from regional climate models that are nested within general circulation models. Spatial variability in rainfall within the Edwards Aquifer region, and the variations in pumping patterns, both of which will impact spring flows, should also be explored in scenario investigations.

Past droughts of shorter duration with more or less intensity are also of interest in understanding the effectiveness of flow protection measures and to test the model's accuracy. A review of PDSI records available for the NOAA Climate Divisions 6 and 7 in Texas covering the Edwards Aquifer region clearly shows the occurrence of such less severe droughts (see Figure 2-2).

Setting aside the drought of record (1951-1956), there are many other periods when the severity of drought was in the categories of mild or more severe; all of them are shorter in duration than the drought of record, generally two to four years. The following periods can be identified from Figure 2-2: 1909-1911, 1933-1934, 1962-1963, 1988-1989, and 2011-2014. The effect of the shorter drought periods can be seen clearly in the observed spring flow record at Comal Springs, and during many of the droughts, the flow decreased below 100 cfs (see Figure 2-3). Consequently, they represent lesser extremes but were severe enough to cause a significant decrease in spring flows. Testing how well the model can predict responses during such lesser extremes may demonstrate its applicability to a variety of climatic conditions and further enhance the confidence in the model for adaptive management and for other applications in Phase 2 of the HCP.

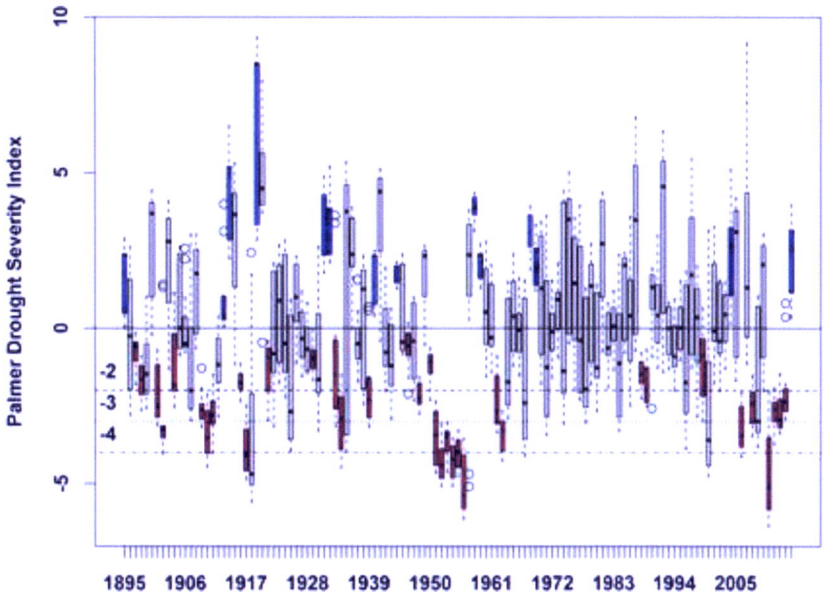

FIGURE 2-2 The Committee compiled and graphed these data of the Monthly Palmer Drought Severity Index (PDSI) for the 1895-2015 period for NCDC Climate Division 6 in Texas. The drought indicators of PDSI= –2, –3, and –4 are noted as dashed lines. The shaded boxes show wet (blue) and dry (brown) years. The drought severity scale is (a) –1.0 to –2.0 = mild drought; (b) –2.0 to –3.0 = moderate drought; (c) –3.0 to –4.0 = severe drought; and (d) greater than –4.0 = extreme drought. Developed by the Committee using the data from: https://www1.ncdc.noaa.gov/pub/data/cirs/climdiv/.
NCDC = National Climate Data Center.

Optimize the Bottom-Up Package

Another useful scenario for the hydrologic modelers to run can answer the question "Can implementation of the four spring flow protection measures of the HCP be optimized?" In the HCP, the four measures in the bottom-up package were used in an incremental manner, buiding the package by superimposing one measure onto another previously implemented measure. For example, since VISPO alone is not adequate in achieving spring flow targets, a combined package of VISPO and the RWCP was tested. It was determined that all four measures were necessary to achieve spring flow targets under the conditions of the drought of record with all permitted withdrawals. All measures were accounted for in the model by making changes to water withdrawals. There is no information on any at-

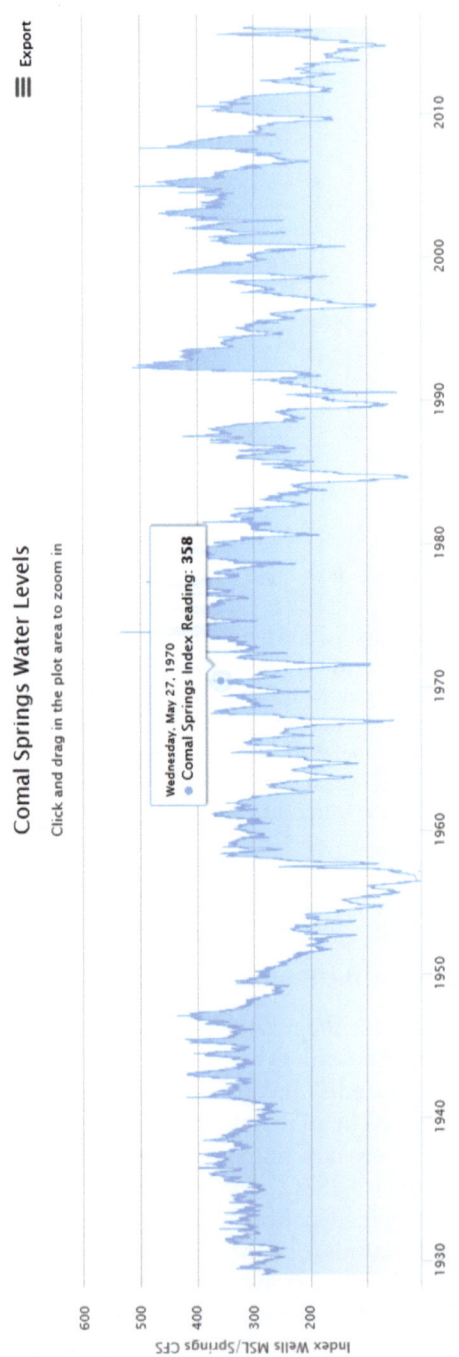

FIGURE 2-3 Observed flows at Comal Spring.
SOURCE: www.edwardsaquifer.org/dataflow/api/chart.

tempt to optimize the combination of measures, including the magnitude and spatial implementation of each or the order in which they might be implemented. The Committee recommends that the EAA undertake an optimization analysis of various combinations of the bottom-up package. In such an analysis, the objective function could be formulated to minimize the deviations of the spring flow and water level targets. From this exercise a different combination of measures with different magnitudes may emerge as the optimal combination that minimizes the deviations from the spring flow targets or cost of implementation.

Understanding the relative effectiveness of various flow protection measures may prove to be extremely valuable for adaptive management and potential revisions of the bottom-up package in Phase 2. Not all droughts will be as severe as the drought of record and, as discussed in a previous section, there will be many more droughts that are less severe. Some of these droughts will require implementation of flow protection measures. Depending on the magnitude of the drought, a particular measure alone may provide the level of protection needed to maintain spring flows necessary to achieve the biological goals of the HCP. For instance, in certain situations, using only the ASR option, which appears to have the greatest "lift" in terms of improving spring flows, may be adequate. The optimization exercises described above will provide the necessary information for decision making either in an adaptive management setting or for revisions of flow protection measures that may be necessary in Phase 2 of the HCP.

Influence of Spatial Pattern of Pumping

The groundwater model includes pumping from a large number of wells and well fields within its model domain that are spatially distributed in a non-uniform manner. In general, certain wells or group of wells may have a larger influence on spring flows than others that are located remotely with respect to the locations of the springs. A comprehensive analysis of this could provide useful information for developing various options for implementing flow protection measures during future droughts. This scenario can answer the question "Which wells have the greatest influence on index wells or discharges from the springs?"

Such a sensitivity analysis involves conducting field tests using a set of wells thought to have the highest sensitivity to water levels at index wells and flows at springs. Pumping at these wells could be increased by some percentage for a certain length of time (e.g., one-two months); with careful monitoring, the data from such a field test could provide valuable information for further validation of the model. Ideally, such a test should be conducted during a period in which the withdrawals have the largest influence and the other stressors (e.g., rainfall) are minimal. Consequently, the ideal time for such a test should be selected carefully. The wells belonging to a large permit holder (e.g., the San Antonio Water System) may have to be used for this testing since it would be difficult to facilitate the involvement of a large number of individual well owners.

The sensitivity analysis should be followed by an optimization modeling exercise to determine the combination of wells and well fields that would be most effective in achieving the hydrologic goals of the HCP. This analysis may be conducted for droughts of various magnitudes. The optimization package may include the contraints due to, say, water rights of certain users. The groundwater management package developed by the USGS is an appropriate tool for optimization analysis (see Box 2-2).

BOX 2-2
A Groundwater Management Optimization Tool for MODFLOW

To best manage groundwater availability in the Edwards Aquifer for optimum habitat protection at spring outlets, the EAA could benefit from a groundwater management tool that can be applied to the MODFLOW model. The Groundwater Management Process for MODFLOW (GWM) is a free, open-source software application available from the USGS (Banta and Ahlfeld, 2013; http://water.usgs.gov/ogw/gwm/). The GWM is an optimization tool that implements decision variables to minimize or maximize an objective function under a set of defined constraints. The objective function represents the costs or benefits resulting from water-management decisions, in terms of economic or societal value. Decision variables include pumping rates from wells, transfer of water to an external reservoir (e.g., ASR), and anthropogenic groundwater recharge. Binary decision variables also are allowed when, for example, a pumping well is either operational or not operational. Constraints can be placed on decision variables (e.g., maximum number of operating wells), hydraulic heads at specified locations and times, and stream flows calculated by the Streamflow-Routing Package (STR1) for MODFLOW (Prudic et al., 2004). STR1 can be used to simulate spring flow as well as stream flow.

By applying GWM to the EAA's MODFLOW model, minimum spring flow thresholds and groundwater levels could be better managed, as influenced by the rate and timing of groundwater withdrawal, ASR, and anthropogenic recharge. For example, a groundwater management problem could consist of decisions about pumping rates and pumping periods for multiple wells to help maintain minimum flow rates from Comal and San Marcos Springs. GWM also could be used to optimize management of ASR. These analyses could be applied under different climatic scenarios, such as a simulated drought.

Significant Growth and Land-use Change in the Recharge Area

The Edwards Aquifer region encompassing as many as 12 counties in South Central Texas is located in one of the fastest-growing regions in the country. During the decade of 2000-2010, the population in the region increased by about 20 percent (Lal et al., 2012). Over the next two to three decades, further growth is expected. It is estimated that between 2010 and 2040 as much as 240,000 acres of available undeveloped land will be converted to developed land (County of Bexar, 2015). The projected land-use change will most likely result in conversion of agricultural land to urban growth, and such conversions typically have a significant impact on rainfall-runoff-recharge processes in a basin. According to Lal et al. (2012), there has been a net reduction of about 130,000 acres of total farmland between the years 2002 and 2007 alone. Urbanization typically results in rapid runoff with decreasing opportunities for recharge.

Since recharge is one of the most important components of the water budget, any change in its characteristics due to land-use changes in the region, and in particular over the contributing zone, has the potential to impact spring flow characteristics. Another complicating factor that would negatively impact recharge quantity is climate change. Projected warming and potentially drier conditions in the basin may lead to less recharge.

A scenario with projected land-use changes and likely change in climate (but no change in water withdrawals by well pumping) over the next two to three decades should be simulated to answer the question "How would a changes in recharge amount due to changing land use impact spring flows?" It should be noted that the current empirical method of estimating recharge calibrated using historical data (i.e., the Puente method) may not allow an assessment of the impact of land-use changes. One of the more physically based models of recharge previously discussed, such as HSPF or the soil-water-balance model by Westenbroek et al. (2010), will be required for such a scenario investigation.

MODEL MANAGEMENT: USING THE MODEL IN MAINTENANCE MODE

Because the EAA's groundwater management model (MODFLOW model) will be used for long-term planning, it is most useful if updated and improved periodically. Improvements may be the result of the availability of additional observations and other supporting data, updates to the conceptual model and hydrogeologic framework, improved versions of the model code, and better parameter and uncertainty estimation methods. The Committee recommends further improvement and model testing to prepare the model for maintenance mode.

Versioning

Once the model moves from the development and calibration stage to operational mode, it should be formally documented as a public record at a high level of transparency. Each periodic model update should be formalized and documented in a peer-reviewed report as a citable model version. Rigorous model description leads to more effective future use of the model, particularly as EAA personnel changes over time. The Committee recommends the use of a formal versioning system, consisting of a model archive and peer-reviewed report identified by a unique version number, with a model update occurring about every five years. At a minimum, model updates would include an assessment of the model's skill in simulating the additional five years of spring flow and water-level records, possibly including recalibration. This assessment should be described in the report and used to track the model's predictive skill with each successive version; hopefully this skill would improve over time. The model archive should be made available to other government agencies and interested parties for the purpose of simulating specific scenarios of interest or for confirming the EAA's published simulation results.

Decision Support System

To ensure minimum continuous spring flows, the HCP specifies flow protection measures, some of which are triggered at specific groundwater elevations at selected index wells. For example, the Stage V Critical Management Period pumping reductions of 44 percent are triggered at 625 feet mean sea level (MSL) at well J-17 and 840 feet MSL at well J-27. In planning for Phase 2 of the HCP, consideration should be given to developing a more refined framework that incorporates modeling into the decision criteria rather than relying on triggers based on measured groundwater elevations at specific wells. Hence, the Committee recommends the development of a decision support system (DSS) to be used in Phase 2 of the HCP in order to apply the model to short-term decisions (e.g., a one-month time frame). This is necessary because short-term decisions that should be made quickly might be substantially delayed if a DSS is not in place as an objective guide. A DSS would clearly direct these decisions on the basis of different model outcomes. A good DSS is developed and applied with the understanding that model predictions, although uncertain, represent the best available science on which to base management decisions.

The DSS should include a protocol for continually incorporating real-time data (e.g., groundwater levels, rainfall, and well withdrawals) and for scheduling frequent model simulations to predict water levels and spring flows for the short-term future. The next step in developing the DSS would be to define the actions to be taken on the basis of an agreed upon probability that a particular outcome will occur. For example, the 12-month outlook of the water levels at an index well would be presented probabilistically, and a pre-determined action would be taken if there is reasonable probability that the water level will be at or below a critical value within that 12-month period. This way, early management actions can alleviate probable undesirable outcomes later. Such a tool would be even more valuable if future climate outlooks are incorporated into the probabilistic predictions. An example of such an approach, known as Position Analysis, is described in Box 2-3. In this case, the approach applies a range of probabilistic meteorological conditions for one- to two-year predictions of Lake Okeechobee water levels.

CONCLUSIONS AND RECOMMENDATIONS

Although a number of improvements to the groundwater model calibration have been achieved since the 2015 NRC report, continued model development would improve the reliability and predictive capability of the model. Particularly useful would be to evaluate the sensitivity of the model to the various scenarios described herein and to test the model's ability to predict spring flow and well levels under recent climatic conditions.

The EAA is encouraged to incorporate additional recommendations from the first Committee report, such as more extensive uncertainty analysis, testing conceptual models on subgrids, and better documentation of model updates (including incorporation of new field data). More extensive uncertainty analysis will enhance confidence in the model results. Telescoping grids can be used to test conceptual models using a smaller area, which can address questions that are difficult to answer with the larger grid. In addition to documenting changes in the model, updating model parameters and input data more frequently assures that the latest information is used to make model predictions.

The groundwater model should be tested against the 2011 to 2015 period, which was not used in model calibration. This period, which includes both very dry and wet years, offers a remarkable opportunity to validate the model and enhance confidence in the model for future applications. Testing the model using the 2011-2015 period is likely to reveal the limitations of the current model. In addition, it should provide information on relative effects of withdrawals and effectiveness of management measures that were implemented during this period. The hydrologic, climatic, and well withdrawal data and the information on management actions for 2011-2015 should be more accurate than those from prior years, allowing for a more reliable assessment of the model.

Several scenarios are suggested for the hydrologic model, including optimizing the bottom-up package, evaluating spatial variations in pump-

> **BOX 2-3**
> **Position Analysis of Large Water Resources Systems**
>
> Position Analysis (Hirsh, 1978; Smith et al., 1992; Tasker and Dunne, 1997; Cadavid et al., 1999) is a form of risk analysis that can assess future risks associated with specific operational plans for a basin over a period of several months, given the current state of the hydrologic system. It relies on the simulation of a large number of possible traces of climatological inputs to the system using the current conditions as the initial values for modeling. To be most useful, Position Analysis needs to incorporate the broadest range of meteorological conditions that may occur in the future, but cannot be selectively forecast.
>
> Using hydrological simulation models, the South Florida Water Management District (www.sfwmd.gov) routinely performs a Position Analysis that produces quantile graphics for several significant water bodies, canals, and gauge locations (Figure 2-3-1 shown for Lake Okeechobee). The lines can also be called "iso-percentile lines." These graphics represent a statistical summary of the simulated stages for a given location. They provide the probability of the stage being below a given value, for every day of the year, based on a current initial stage and the rainfall regime experienced by that feature each year for the available simulation period, running 365 days from initialization. For instance, for all the stages shown on the 80% line, the probability of being below that stage is 80%, while the probability of being above is 20%. The 50th percentile is the median stage each day, thus half the years on that day were above that value and half were below. One should not expect that a given iso-percentile line comes from a single simulated year. They are usually formed with values coming from different years. This provides a useful probabilistic indication of where the stage level could go. It is reasonable to accept that above-average rainfall at a given location will lead to higher than median stages in that area, but there is no one-to-one relationship between rainfall and the stage values. Other factors are involved, not least of which are the management criteria for moving water through the system.

ing, and predicting how significant growth and land-use change in the recharge area might affect spring flows. Testing a variety of scenarios will not only improve the confidence in the model itself but will also help develop strategic decisions associated with adaptive management and revisions to minimization and mitigation measures.

The Five-Year plan for the hydrologic model should include formal versioning and a decision support system that will be useful in future phases of HCP. The model should be updated every five years, with each new version including a peer-reviewed report and permanent archive of the numerical

HYDROLOGIC MODELING

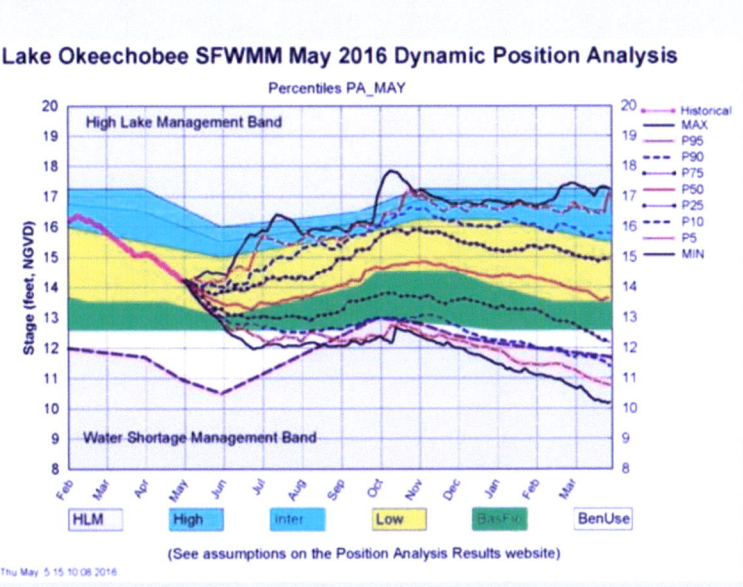

FIGURE 2-3-1 Position Analysis output of the water levels in Lake Okeechobee in South Florida for the month of May 2016. The percentiles shown are generated by running a complex system simulation model for many years of historical hydrology, one year at a time with the system initialized (both surface water and groundwater) to the current conditions corresponding to the beginning of May 2016. The initial conditions account for antecedent conditions that correspond to months prior to May. This analysis is produced at the beginning of every month to aid in decision making in terms of management actions that could avoid future undesirable states of the system (e.g., very low lake levels).

model that is available to the public. A decision support system will help minimize the subjectivity of management decisions that require a rapid response and should be included in Phase 2 of the HCP.

REFERENCES

Anderson, M. P., W. W. Woessner, and R. J. Hunt. 2015. Applied Groundwater Modeling: Simulation of Flow and Advective Transport. Academic Press.

Banta, E. R., and D. P. Ahlfeld. 2013. GWM-VI—Groundwater Management with Parallel Processing for Multiple MODFLOW versions: U.S. Geological Survey Techniques and Methods 6-A48.

Brakefield, L. K., J. T. White, N. A. Houston, and J. V. Thomas. 2015. Updated numerical model with uncertainty assessment of 1950–56 drought conditions on brackish-water movement within the Edwards Aquifer, San Antonio, Texas. U.S. Geological Survey Scientific Investigations Report 2015–5081, 54 pp., http://dx.doi.org/10.3133/sir20155081.

Cadavid, L. G., R. Van Zee, C. White, P. Trimble, and J. T. B. Obeysekera. 1999. Operational Hydrology in South Florida Using Climate Forecast. American Geophysical Union, Proceedings of the Nineteenth Annual Hydrology Days, August 16-20.

County of Bexar. 2015. Southern Edwards Plateau, Habitat Conservation Plan Prepared by Bowman Consulting Group, LTD 310 Bee Cave Road, suite 1000, Austin, Texas 78746. Project Number 005520-01-001.

EAA. 2015a. National Academy of Sciences—Review of the Edwards Aquifer Habitat Conservation Plan. Report 1 Implementation Plan. EAA August 20, 2015.

EAA. 2015b. Hydrologic Data Report for 2014. Report No. 15-01, November 2015.

EARIP. 2012. Habitat Conservation Plan. Edwards Aquifer Recovery Implementation Program.

Enzenhoefer, R., T. Bunk, and W. Nowak. 2014. Nine steps to risk-informed wellhead protection and management: A case study. Groundwater 52(S1):161-174.

Fratesi, S. B., R. T. Green, F. P. Bertetti, R. N. McGinnis, N. Toll, H. Başağaoğlu, L. Gergen, J. R. Winterlee, Y. Cabeza, and J. Carrera. 2015. Final Report: Development of a Finite-Element Method Groundwater Flow Model for the Edwards Aquifer. SWRI Project No. 20-17344

Hirsch, R. M. 1978. Risk Analysis for a Water-Supply System—Occoquan Reservoir, Fairfax and Prince William counties, Virginia. Hydrologic Science Bulletin 23(4):476-505.

Lal, R., B. A. Stewart, V. Uddameri, and V. P. Singh. 2012. Competition between Environmental, Urban, and Rural Groundwater Demands and the Impacts on Agriculture in Edwards Aquifer Area, Texas. Pp. 117-130 In: *Soil Water and Agronomic Productivity*. CRC Press.

Loáiciga H. A., J. B. Valdes, R. Vogel, J. Garvey, and H. H. Schwarz. 1996. Global warming and the hydrologic cycle. Journal of Hydrology 174 (1 and 2):83-128.

Mace, R. E., and S. C. Wade. 2008. In hot water? How climate change may (or may not) affect the groundwater resources of Texas. Gulf Coast Association of Geological Societies Transactions 58:655-668.

NASEM (National Academies of Sciences, Engineering, and Medicine). 2016. Evaluation of the Predictive Ecological Model for the Edwards Aquifer Habitat Conservation Plan: An Interim Report as Part of Phase 2. Washington, DC: The National Academies Press. doi: 10.17226.23577.

NRC (National Research Council). 2015. Review of the Edwards Aquifer Habitat Conservation Plan: Report 1. Washington, DC: The National Academies Press.

Pappenberger, F., and K. J. Beven. 2006. Ignorance is bliss: Or seven reasons not to use uncertainty analysis. Water Resources Research 42(5).

Prudic, D. E., L. F. Konikow, and E. R. Banta. 2004. A new Streamflow-Routing (SFR1) Package to simulate stream-aquifer interaction with MODFLOW-2000: U.S. Geological Survey Open-File Report 2004-1042, 95 pp.

Puente, C. 1978. Method of Estimating Natural Recharge to the Edwards aquifer in the San Antonio area, Texas. U.S. Geological Survey Water-Resources Investigations 78-10, 34.

Smith, J. A., G. N. Day, and M. D. Kane. 1992. Nonparametric framework for long-range streamflow forecasting. Journal of Water Resources Planning and Management, ASCE 118(1), 82-92.

Tasker, G. D., and P. D. Dunne. 1997. Bootstrap position analysis for forecasting low flow frequency. Journal of Water Resources Planning and Management ASCE 123(6):359-367.

Westenbroek, S. M., V. A. Kelson, W. R. Dripps, R. J. Hunt, and K. R. Bradbury. 2010. SWB—A modified Thornthwaite-Mather soil-water-balance code for estimating groundwater recharge. U.S. Geological Survey Techniques and Methods 6-A31, 60 p.

Winterlee, J. 2014. MODFLOW Verification Analysis pptx file. Sent to the NRC Committee on 2/14/2014.

Winterlee, J. 2015. EAA Modeling Program Update. Presentation to the National Academies' Committee on 10/29/2015.

Winterlee, J. 2016. EAA Modeling Program Update. Presentation to the National Academies' Committee on 2/3/2016.

3

Ecological Modeling

One of the major efforts set forth by the Habitat Conservation Plan (HCP) is the creation of ecological models for the Comal and San Marcos systems. The HCP describes the overall goals of the modeling effort: "The EAA will . . . develop a predictive ecological model to evaluate potential adverse ecological effects from Covered Activities and to the extent that such effects are determined to occur, to quantify their magnitude. The model results will help the Applicants develop alternative approaches or possible mitigation strategies, if necessary." The ecological models should be able (1) "to predict specific ecological responses of the Comal and San Marcos Springs/River ecosystems and associated Covered Species to various environmental factors, both natural and anthropogenic"; (2) "to assist in establishing potential threshold levels for these ecosystems and associated species relative to potential environmental stressors"; and (3) "to assist the overall scientific effort to better understand the interrelationships among the various factors affecting the dynamics of these ecosystems and associated species." The models are also expected to be able to account for impacts to the ecosystems from both management measures and natural variations, including such things as groundwater withdrawal, recreation activities, parasitism, and restoration actions. The HCP later describes several other structural and operating requirements for the models, but does not go so far as to prescribe exactly which listed species should be included and what processes should be encompassed, noting only that the models should be capable of including plant, animal, hydrological, climatic, and management variables, and simulating interactions among all of these components. In response to the HCP, the Edwards

Aquifer Authority (EAA) created an ecological modeling team consisting of academics, government scientists, consulting firms, EAA employees, and others to develop the first version of the models by December 2016. They focused on the population dynamics of the fountain darter (FD) and the spatial and productivity dynamics of key submersed aquatic vegetation (SAV) species.

The Committee has reviewed the progress made on the ecological modeling twice prior to this report. NRC (2015) discussed the basic design of the FD model, including the decision to develop an individual-based model, and it opined on several precursors to the model, such as the habitat suitability analyses done for FD, Texas wild rice, and the Comal Springs riffle beetle (CSRB). NASEM (2016) reviewed the first complete report from the ecological modeling team on what is now expected to be the sole product—models that predict the abundance of SAV and FD, each run separately and also run in a coupled mode. This chapter has two goals: (1) to address the EAA's responses to NRC (2015), and (2) to suggest scenarios for the FD model to run, now that a calibrated version is available. NASEM (2016) is provided as Appendix A to this report, and some of the recommendations made in that report are summarized in Box 3-1. It is expected that the reader will be knowledgeable about the contents of NASEM (2016) prior to reading this chapter. The comments and suggestions for scenarios presume that the recommendations in NASEM (2016) have been sufficiently addressed.

EAA RESPONSE TO THE COMMITTEE'S FIRST REPORT

Recommendation for Development of a Conceptual Model

In NRC (2015), the Committee "recommended that as a top priority the EAA develop an ecosystem-based conceptual model, or a series of conceptual models of increasing resolution, that show how water quality and quantity, other biota, and restoration and mitigation activities are expected to interact with the indicator species, as well as with all covered species. Boxes in the conceptual model would represent targets of the monitoring program, while arrows linking the boxes would represent quantitative or empirically derived relationships between the boxes based on research. Such interactions for which too little data are available to establish empirical relationships could be targeted for monitoring and further research during the permit period." Also, with respect to project integration, NRC (2015) states that "the HCP would benefit from more formal integration to enable clear explanation of the many sets of results emanating from the monitoring, modeling, and research efforts. Without greater attention to project integration, there is a danger that the large number of separate projects will not combine seamlessly into an overall science program." An overall con-

> **BOX 3-1**
> **Recommendations Related to Improving the**
> **Fountain Darter Ecological Model**
>
> - A simple one-time transfer of the models from the developers to the EAA should be avoided because this can result in inefficient, and even possibly erroneous, use of the FD and SAV models.
> - The focus on using the FD model to predict the responses of FD abundance to alternative HCP flow control packages is useful, but there are other uses of such mechanistic models that should be considered.
> - The temporal and spatial scales of the SAV and FD models are reasonable but the representativeness of selected reaches and the variance properties associated with the use of QUAL2E outputs as model inputs should be clearly documented.
> - The use of an individual-based approach imbedded within a 2-D spatial grid for full life-cycle simulations of FD population dynamics is a scientifically sound framework for the questions being asked, but there remain some important steps (related to how SAV is represented) to link the FD dynamics to their habitat.
> - The representation of the processes of FD growth, mortality, reproduction, and movement presently in the model are well-founded but may be too simple and not sufficiently linked to changes in habitat and flow to answer some of the important management questions.
> - Thresholds in process representations should be used cautiously because they can erroneously create nonlinear population responses and unrealistic sensitivities to changes in habitat and flow.
> - The representation of density-dependence and how its effects on individuals manifest at the population level needs further evaluation.
> - The representation of flow effects in the model seems too limited in potential effects due to reliance on having site-specific empirical evidence for the effects.
> - Calibration and validation of the FD model to date show the model can reproduce the historical abundances, but additional confidence is needed to most effectively use the model for management purposes.
> - The historical time period used for calibration had relatively similar environmental conditions from year-to-year, which limits the range of conditions of scenarios feasible for exploration by the model.
>
> SOURCE: NASEM (2016).

ceptual model of the system including hydrologic, climate, and biological components was identified as critical to such integration.

The EAA has now provided a scientifically sound foundation on the way to developing a generalized ecosystem-based conceptual model. The process of developing the FD and SAV models, and the associated conceptual diagrams of how the models work, provide an excellent basis for further development of an overall conceptual model. While the conceptual diagram from the models is not as comprehensive as the Committee suggested in the first report, it is a major improvement over the original influence diagrams of the HCP and reflects well the current level of understanding (related to FD and SAV) in the Edwards Aquifer system. As shown in Figure 3-1, the conceptual models show linkages between potential forcing factors (e.g., spring flows and water quality) and important response variables (SAV and FD abundances). Additionally, the EAA has adopted the multidimensional surface water modeling system (MD-SWMS) for modeling surface water dynamics, the stream water quality model QUAL2E for modeling water quality, and generated submodels for SAV (see Figure 3-1) and FD. We encourage EAA to continue with the conceptualization of the overall ecosystem by building on the FD and SAV conceptual models.

It is hoped that the conceptual models produced to date, and their further expansion to the overall ecosystem, will not only serve to guide development of the predictive models, but will provide a powerful integrative communication tool for the overall HCP and better coordinate the diverse expertise found across EAA's multiple advisory committees and contractors, particularly in cases where differences in opinion, interpretation, and understanding might be prevalent. In addition, the conceptual and predictive ecological models should be used to evaluate the minimization and mitigation (M&M) measures, in terms of both appropriateness and efficacy. For example, the HCP (on pages 4-43 through 4-45) hypothesizes that M&M measures will have important impacts on habitat and population sizes for FDs, CSRB, and Texas wild rice. The conceptual models can help devise priorities for M&M measures, while measured impacts of the M&M measures can be used to fine-tune the predictive models. As described in NASEM (2016), the progression through model development, testing, and usage is iterative. Thus, as Phase 1 of the HCP progresses, it is expected that M&M priorities, as well as the conceptual and predictive models, will continually improve as new data are collected and incorporated.

Recommendations about Habitat Suitability Analyses

NRC (2015) suggested that, given the absence of an ecological model for Texas wild rice, the current habitat suitability analysis (Hardy et al., 2010) should be treated as a hypothesis and tested for robustness throughout the San Marcos River. For example, the M&M activities could be used to test the validity of using water depth and velocity as the only predictive variables for optimal habitat for Texas wild rice. The Recommendations Review Work Group (RRWG) responded that they were working on this

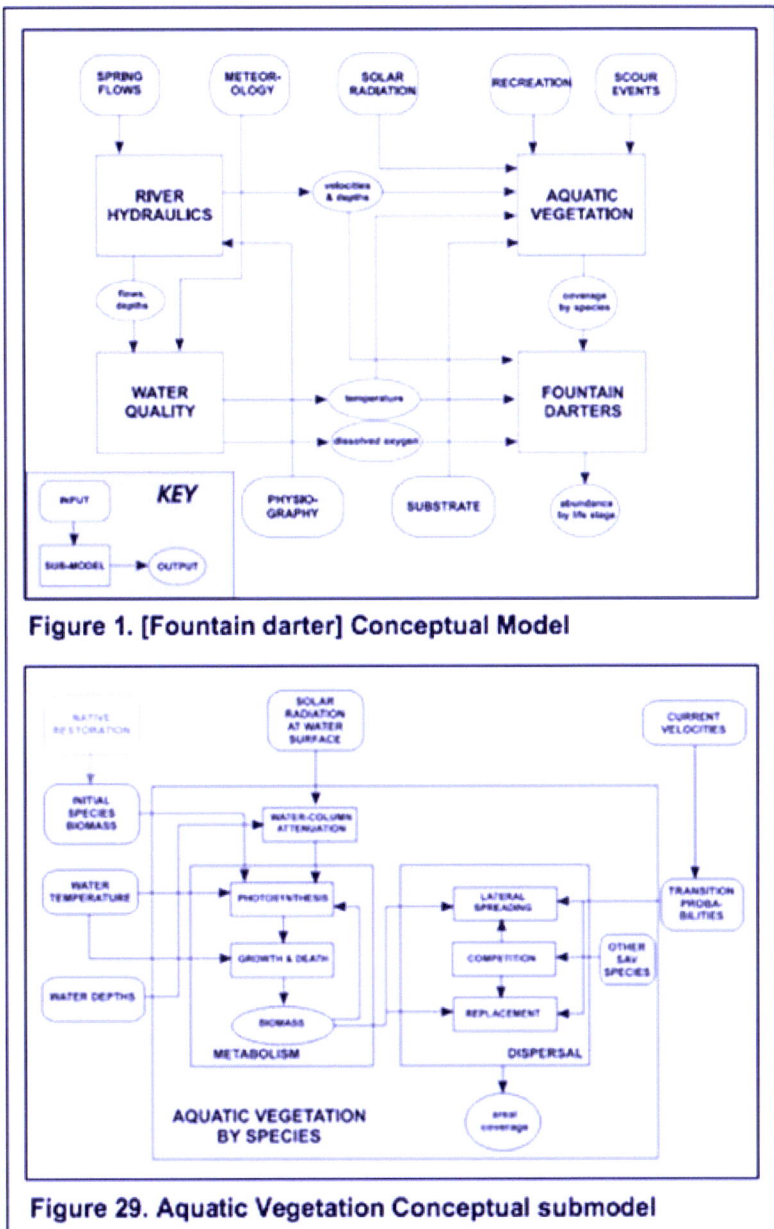

FIGURE 3-1 Generalized frameworks developed by the EAA providing conceptual linkages between potential forcing factors and important response variables for fountain darter and submersed aquatic vegetation abundances.
SOURCE: Adapted from BIO-WEST, 2015.

("Continual"), suggesting that the continued replanting of Texas wild rice will be conducted with such tests in mind.

Similarly, NRC (2015) recommended that the habitat suitability analyses done for the FD act as a "back-up" to the individual-based modeling and provide additional quasi-independent results to support a weight-of-evidence approach for FD. The RRWG's response ("Done") made it clear that they did not see the value of continuing work on the habitat suitability analysis for FD given their focus on developing the mechanistic ecological model. Nonetheless, they acknowledged that "if the fountain darter module fails or does not calibrate, then suitability should be revisited" (EAA, 2015).

If the suitability analyses are pursued in the future, the EAA should return to NRC (2015) for a thorough evaluation and recommendations on their earlier approach and consider new methods that have evolved to address some of the issues with the classical habitat suitability approach (Guisan et al., 2013; Merow et al., 2014; Hamilton et al., 2015). In particular, such analyses should be based on careful selection of spatial scales, and it is important that spatial and temporal resolution are aligned throughout all sources of data. Furthermore, parameters and the estimation of functional relationships, as well as evaluation of alternative model formulations, should be based on sound statistical metrics.

Recommendations about Comal Springs Riffle Beetle

NRC (2015) stated that prior to being able to include the CSRB in a mechanistic model, it is critical to have a much deeper understanding of the spatial distribution, range of potential habitats, and natural history of the CSRB. This natural history includes understanding the number of generations per year, cohort synchrony or asynchrony, the times of year for reproduction, and the biotic and abiotic variables that influence these dynamics (e.g., siltation). Furthermore, a better understanding of the optimal CSRB habitat is needed to understand how changing flow conditions will impact CSRB. The RRWG responded positively to this item, devoting the 2016 Applied Research budget exclusively to CSRB research projects. There are also planned to be at least two Applied Research projects in 2017 devoted to the CSRB (see Chapter 5).

It is unlikely that an ecological model of the CSRB will be developed in the near future. Regardless, the data being collected on the beetle are potentially of great importance, warranting a few comments about the most recent studies on CSRB abundance in the Comal system. In 2014, as part of the Applied Research Program, the EAA contracted with Zara Environmental for a system-wide estimate of the CSRB population within the Comal Springs ecosystem. As described in detail in Chapter 5, this was the most directed research effort to date for estimating CSRB population abundance. Nonetheless, there were several serious flaws to the study design, including the sampling approach. Furthermore, results from the final report (Zara Environmental, 2015) suggest that the CSRB populations are very low compared to previous reports that estimated CSRB populations based on the wetted area of potential habitat. The discrepancies between these various population estimates illuminate why a better sampling approach is necessary for estimating the current CSRB population and projecting future changes. Because the long-term biological goals for the CSRB involve both a "qualitative habitat component and a quantitative population measurement" (page 4-9, EARIP, 2012), a better sampling method is also critical to determining compliance with the Incidental Take Permit.

The HCP suggests that the CSRB may be an indicator species for evaluating the impact of covered activities on other listed species. That is, page 4-38 of the HCP states that "In 2010, the EARIP held workshops involving a multi-disciplinary team of biologists to develop influence diagrams regarding the impacts on fountain darters, Texas wild rice, and the Comal Springs riffle beetle. These species were believed to be good indicator species for the impacts on other Covered Species." There appears to be disagreement about whether the CSRB is an indicator species, particularly because assumptions about CSRB behavior during low flows (retreating into subterranean habitat) may not hold for other invertebrates or amphibians. If the CSRB is abandoned as an indicator species, the EAA should be prepared to develop detailed monitoring plans for the other covered species (e.g., dryopid beetles, Peck's Cave amphipod, salamanders). Planned Applied Research projects suggest that the EAA is moving in this direction (see Chapter 5).

SCENARIOS FOR ECOLOGICAL MODELING

As part of its statement of task, the Committee was asked to "identify those biological and hydrological questions related to achieving compliance with the HCP's biological goals and objectives that the ecological and hydrologic models should be used to answer, specifically including which scenarios to run in the models." EAA has focused the ecological modeling efforts on FD and SAV. Recommendations from the Committee's June report on the ecological models (Box 3-1, NASEM, 2016) suggest that, given the present state of development of the models, it will be difficult to run scenarios of interest to the EAA. For example, NASEM (2016) states that "the representation of fountain darter growth, mortality, reproduction, and movement may be too simple and not sufficiently linked to changes in habitat and flow to answer some of the important management questions." "For both fountain darter and SAV, the representation of flow effects in the model is too limited because of reliance on having site-specific empirical evidence for the effects." Furthermore, "the historical time period used for calibration of the fountain darter submodel had relatively similar environmental conditions from year to year, which limits the range of conditions of scenarios feasible for exploration by the model."

Despite these remaining model development and testing steps, this chapter explores the development of scenarios for the ecological models, focusing on the FD model because it is further along in development than the SAV model. Only the most general guidelines for scenarios to run in the SAV model are possible at this time due to the early nature of that modeling effort. The reader is reminded that the issues and recommendations described in the Committee's previous report (NASEM, 2016, which is Appendix A in this report) should be adequately addressed prior to running the scenarios. Indeed, the manner and degree to which these issues are addressed will determine which scenarios can be run and the confidence level appropriate for interpreting the results of the scenario analyses.

While the following section is largely directed toward the FD model, there is an important relationship between the two modeling efforts. Many of the management actions related to FD will necessarily involve management actions directed at SAV (as FD habitat). Explicit treatment of how actions directed at SAV would affect FD through the coupled models is preferred. However, such explicit analysis requires that the two modeling efforts progress sufficiently to allow them to be coupled. The capability to couple two mature and tested models (FD and SAV) will enable more questions to be addressed (e.g., dynamic and simultaneous responses of FD and SAV to changes in flow) and will make the predictions of SAV effects on FD more defensible.

To varying degrees, the scenarios described below require that the recommendations of the NRC (2016) report (Box 3-1) be adequately addressed. The degree to which the recommendations are addressed will determine the confidence and credibility of the model predictions for many of the scenarios. First, a set of concepts about best practices in designing and interpreting scenarios is described, adapted from Rose et al. (2015); Addison et al. (2013) also offer useful advice on using ecological models for management analyses. The ideas and concepts for designing scenarios apply to both the FD and SAV models, as well to ecological models in general. Then, a set of possible scenarios specific to the FD model are provided to illustrate the types of questions that could be addressed once the model is deemed management-ready. Much discussion about scenario analyses comes from the business community (e.g., Bradfield et al., 2005) and from climate change modeling (Parson et al., 2007; Lempert, 2013), which focuses on possible future conditions.

Concepts for Designing Ecological Model Scenarios

Framing of Scenario Analyses

The terms "scenario analysis," "sensitivity analysis," "uncertainty analysis," and "model experiments" are widely used, often interchangeably, in simulation modeling. Similarly, the results of such analyses are called "predictions," "projections," and "forecasts." These words can mean different things to different people, and can create confusion among modelers and end-users leading to miscommunication, improper interpretation of results, and unachievable expectations placed on model products.

Scenario analysis, sensitivity analysis, uncertainty analysis, and model experiments all involve changing model inputs to assess how the model responds. A "scenario" is a coherent, internally consistent and plausible description of a possible future state (http://www.ipcc-data.org/guidelines/pages/definitions.html). Model experiments are where key conditions (which may not be observable in nature) are used in various combinations to perform experiments (e.g., to test hypotheses, experimental design) (Peck, 2004). Sensitivity analysis usually involves the small variation of individual parameters or other inputs, while uncertainty analysis involves realistic and simultaneous variation in model inputs (Saltelli and Annoni, 2010).

Prediction is a general term regarding model results, with a "projection" being considered a less rigorous prediction than a "forecast." Specific years are typically associated with forecasts, implying the results are what should be expected in nature (e.g., FD abundance in 2021). Projections can be expectations of model predictions over a number of years of calculation, rather than associated with specific dates. Care should be used in whether scenario results are labeled with real years (e.g., 2020, 2021, 2022, etc.) or arbitrary years (10, 11, 12, etc.). The label given to the analyses of management-driven questions using the FD model will be important for clarity and communication.

In addition to properly labeling the model simulations, it is also critical to specify the actual simulations themselves in as much detail as possible to ensure the modeling results will be useful and credible. Often, modeling is considered unsuccessful because of the lack of specification of the questions to be answered, coupled with people having overly high expectations of what the modeling can do (e.g., expect forecasts). To illustrate, a poorly structured question is: (a) What are the effects of low flow on fountain darter? versus the well-stated question (b) How do two consecutive years of 1 percent chance droughts within 20 years of historical conditions of flow affect the long-term (20-year) average annual population abundance of adult fountain darter? Question (a) is vague about the conditions of interest and the response variable and the response variable's time scale of response.

Question (b) provides critical details on what is meant by "low flow" (two consecutive 1/100 year droughts) and "affect FD" (20-year average adult population size). The questions should also be informed by the types and needs of the management actions being considered under the Habitat Conservation Plan.

The Committee recommends that all scenario questions be well defined and the model results carefully labeled. The more specific the question associated with the scenario is stated, the more likely the model can provide an answer.

Domain of Applicability

Models often have many hidden assumptions. A major hidden assumption is about the range of input values over which certain relationships are valid and the labeling of inputs with general names but then using them in very specific ways. These hidden assumptions, along with the range of conditions over which the model is been evaluated (e.g., calibration and validation datasets), define the domain of applicability of the model. Scenarios that push the model outside its domain result in increasingly uncertain predictions.

To illustrate, consider a fish population model that has an input labeled "flow." However, the equation in the model that uses flow was a linear relationship of its effect on mortality rate, and estimated over a narrow range of flow values. Also, because of the previously small variation in flow in the years used to calibrate the population model, the effect of flow on other possible processes (e.g., timing of spawning, growth rate) were reasonably ignored. Thus, simply changing flow in the model to represent very low flow years (e.g., drought) can result in inaccurate predictions. Inaccuracies arise because the new values of flow would actually result in more than a linear change in mortality rate than is assumed in the model (inadequate process representation). Also the effects of flow on spawning and growth were ignored because of the narrow range of variation used when the model was formulated (missing effects).

The conditions under which the model was developed should be compared to the conditions for which the model will be used in scenarios, in order to determine the degree to which the model is within in its domain of applicability. Are the changes and expected effects within the range for which the model has been tested or evaluated? Do the effects approach extreme aspects of the relationships where there is high uncertainty or where responses do not adhere to the assumed relationships?

Explicit Versus Implicit Representation

The changes in factors that are varied as part of scenarios can be represented explicitly or implicitly in the model. Explicit representation means that a variable or factor is named in the model description, and its effects within the model appear in equations. An example would be when the growth rates of fish in a model include a relationship that has flow as an explanatory variable (e.g., growth rates peak at some intermediate flow value). Each day the value of flow is used to determine the growth rate of the individual fish for that day. With such a model, no other changes would be needed to test scenarios about how flow affects growth rates and population dynamics. Different time series of flows can be input to the model, and the predicted population dynamics can be compared.

Implicit representations are when the effect of a factor is imbedded within the formulation of the model, and the factor may not appear on any list of variables or parameters or even anywhere in the model equations. The factor is still included in the model, but its effect is built into the relationships without specifying its effect as a model input. In our simple example, flow would not be explicitly part of the growth rate equation, yet the effects of flow on growth rate are included because whatever growth rates were assumed occurred under some set of flows (typically assumed to be average or representative conditions). To examine model responses to changes in flow requires one to simply assume what changes in the growth rate would occur from a changed flow, and then the model can be run with the original and adjusted growth rates. In fact, if done correctly, one would get the same results from the explicit and implicit representations.

Hence, just examining the list of model variables and parameters or oversimplified diagrams of how the model works is not sufficient to judge the realism of what factors can be changed as part of scenarios. Explicit representations should be scrutinized for how the change in any given factor is represented. Implicit representations do not preclude assessing the effects of a factor, but how the changed conditions were realized by altering existing processes or formulations in the model needs to be evaluated. Overly general phrases like "The effect of low flow was . . ." without a clear explanation of what potential effects were included, and not included, should be avoided. The only way to fully understand what effects are included is to examine the model code itself to see the equations and how they are solved, which is not practical in many situations. Careful documentation that goes beyond general word descriptions and box-arrow diagram descriptions of the model is required so that analysts or knowledgeable staffers can easily respond to questions with specific and accurate answers about what actually was affected by the changed flow conditions. It should be noted that the documentation of the FD model has been excellent to date; it should be noted that all equations, solution methods, and their justification should be documented before they are coded.

Implicit versus explicit representation also applies to spatial and temporal considerations. One does not have to simulate the spatial and temporal scales of every process in order to include their effects in simulations. For example, prey encounters occur on millimeter and second scales, but one does not have to build a model that uses millimeter-sized spatial cells and a one-second time step to include predators' encountering patchily distributed prey. Finer scales than explicitly represented can be assessed implicitly by generating randomness around the function that relates prey to predator consumption or growth (e.g., Letcher and Rice, 1997).

There should be an explanation of the expected effects of a scenario on, for example, fountain darter abundance, and what and how these effects are represented in the model (either explicitly or implicitly). For each scenario, there should be confirmation that the major effects are represented in a reasonable way. For example, if flow is to be varied, then what processes and life stages are expected to be affected?

Uncertainty, Stochasticity, and Variability

Proper interpretation of the results of a model analysis of alternative scenarios depends on how variability is incorporated into predictions. How does one know whether the predicted FD abundances averaged over 10 years are really different among scenarios? We refer to variability as the combined effects of stochasticity and uncertainty. Examples of stochastic effects relevant to the FD model include the occurrence of drought conditions, variation in spring flows from year to year, and fluctuations in abundances of predators. Common sources of uncertainty are the use of laboratory-based measurements to estimate model parameters, use of multiple field studies that occurred in different time periods, and inability to specify unique formulations of processes in the model because alternative formulations result in equally valid fits to the available data. More measurements reduce uncertainty, but not stochasticity (Ferson and Ginzburg, 1996). Appreciating and keeping track of how variability results from uncertainty and stochasticity sources are important when judging the realism of the model and for determining whether differences among alternative scenarios are biologically meaningful.

Observation (or measurement) error is also important to consider when interpreting the results of scenarios. One's confidence and ability to detect differences in predictions is based on the validation of the model using data. Treating the data as having no observation error can result in inaccurate determination of model confidence as part of model validation (Stow et al., 2009), and therefore misinterpretation of the ecological significance of differences among scenarios. For example, when data are treated as exact or overly precise, the model can be expected to generate differences in order to match the data but, in fact, the differences in the data are not reflective of real differences but actually are indistinguishable due to measurement error. This carries over into scenario analysis by putting too much credibility on differences in model predictions across scenarios, when in fact the validation did not support declaring such differences as ecologically meaningful.

Propagating variability through analyses so that final results are ranges or probability distributions should be considered in most all analyses. Purely deterministic analyses (point or simple trend line predictions without uncertainty) do not capture the true variability observed in nature, and there has been much effort to incorporate stochasticity and uncertainty into fish population and food web models to match natural variation (e.g., Bjørkvoll et al., 2012; Link et al. 2012; Magnusson et al., 2013). However, the details of what sources of uncertainty and stochasticity are being considered in the specified variability of the inputs affect how to interpret the spread of results in the output. Saltelli et al. (2004) note that it is rare that an analysis correctly generates realistic variability that is comparable to the observational data; yet, we often interpret the variability of predictions as what is expected in nature. How the variability in predictions of an analysis was generated should be clearly documented, and its implications on how to interpret results should be fully understood.

Critical questions to ask for each scenario include the following. What sources of stochasticity are represented? Is uncertainty kept track of, including uncertainty in the data used to define the scenarios and from the outputs of other models that are used as input to the fountain darter model? How do the predicted differences between scenarios compare to the expected variability that arises from stochasticity and uncertainty? That is, are the differences ecologically significant?

Relative or Absolute Predictions

Model predictions can be divided into two types based on how their predictions are viewed. Some questions require predictions in native units such as annual FD population abundance, while many other scenarios are better viewed as relative predictions. With relative predictions, model predictions are compared to a simulated baseline condition and results expressed as changes from the simulated baseline. These relative predictions are very useful with long-term simulations (future conditions become unceasingly uncertain) because the assumptions of future conditions are maintained in both the baseline and scenarios simulations, and to compare among alternative management options. Although absolute predictions are very tempting because they directly relate to what happens in nature and the model output is labeled as an absolute output, we generally have much more confidence in relative (model-to-model) predictions.

To illustrate, using the FD model to predict whether 40 or 50 or 60 CFS is protective requires a rigorous validation process to determine if the model can predict absolute abundances sufficiently well to distinguish among the 40 to 60 CFS conditions. Similarly, determining if the population abundance (number of individuals) will go below some value and interpreting that prediction as what will occur in nature is tenuous. Use of the same model for relative predictions would express the effects of 40, 50, and 60 CFS as the percent change in simulated FD population abundance from a baseline abundance. The expected benefits of different management actions can be compared to each other very effectively using a mix of absolute prediction viewed as semi-quantitative and relative predictions.

An important consideration with relative predictions (and often also with absolute predictions when, for example, different management actions are to be compared to a no-action alternative) is defining what is baseline. Baseline conditions rarely can be simply defined as some pristine condition because such conditions may be poorly known and undocumented (Pauly, 1995; Papworth et al., 2009) or not achievable due to other changes in the system (Balaguer et al., 2014; Duarte et al., 2015). If comparisons are also needed under future conditions, then determining the baseline becomes even more challenging because the historical or present-day baseline must then be extrapolated to what it would be under future conditions (Higgs et al. 2014).

As part of specifying each scenario, the baseline conditions and dimensions of the predictions (temporal and spatial scales; absolute or relative terms) should be clearly stated.

Explanations for Predicted Results

The power of using ecological models is that, not only can state variable or aggregate predictions (such as population abundance) be made, but the modeling can provide the reasons for the predicted responses. All model results can be explained at the level of the processes represented in the model. If an ecological model, such as the FD model, is well constructed and tested, providing the explanations for predicted responses to scenarios beyond just abundance (e.g., changes in stage survival, fecundity, and spatial distributions) can inform management actions.

All predictions for scenarios should include, at some level, model-based explanations of why the predicted response occurred. For the FD model, this would focus on how growth, mortality, reproduction, and movement differed between baseline and scenario.

Iterative Process

Scenario analysis should be used as part of a broader iterative process inherent in all ecological modeling. The perception that ecological modeling is a linear process (development→ calibration→ validation→ scenarios) diminishes the usefulness of the modeling. The iterative aspects during model conceptualization, development, and testing may not be obvious to outside observers but they occur. Scenarios should be defined based on the management needs, to advance our understanding, and to identify critical data gaps. Often, some scenarios result in the model generating counter-intuitive or unrealistic results. The model may have been pushed beyond its domain of applicability or have assumed relationships for processes that no longer apply, or may have missing processes that only become important under the new scenario-defined conditions. This is a positive result because, once resolved, it strengthens the model structure or helps define what conditions the model can be used to examine for future simulation analyses. Understanding the model's explanation for results can also lead to targeted laboratory and field data collection to improve model formulations and reduce prediction uncertainty (as mentioned in NASEM, 2016, Comment 2).

Example Fountain Darter Model Scenarios

Some example sets of simulations are provided to illustrate the types of questions the FD model could be used to address. These overlap to some degree, and others can be constructed. As mentioned earlier, proper analysis of these scenarios is predicated on the recommendations of NASEM (2016) being addressed. Furthermore, some of these scenarios could be applied to the SAV model once it is further along in development. Finally, some of the scenarios have analogues in Chapter 2, which discusses scenarios for the hydrologic modeling. This is because the two modeling efforts overlap to some extent in their purposes (i.e., both examine aspects of the HCP) and because designing and interpreting scenarios have some commonalities across simulation modeling in general.

Test the Model against Observed Flows

A straightforward scenario would be to use historical flows outside of the calibration and validation time periods to assess FD responses under a wider range of previously observed historical flow conditions (similar to the tests of the hydrologic model mentioned in Chapter 2). This would broaden the domain of applicability of the model.

The Bottom-Up Package of Flow Protection Measures

The effects of the EAA's so-called "bottom-up package" of flow protection measures could be imposed in the model and compared to FD population dynamics without the package (again, similar to what is suggested in Chapter 2 for the hydrologic model). Assumptions about future conditions could then be imposed on both the baseline (future without projects) and bottom-up package scenarios to help guide management actions.

Systematically Vary Flows

The historical record of flows is a limited subset of possible flow patterns that can vary daily, seasonally, and interannually. These levels of variability are overlain on each other to create many possible patterns of flows in 10- or 20-year time periods. Scenarios that systematically vary the daily and seasonal dynamics (when, duration, magnitude), as well as interannual patterns (e.g., occurrences of droughts), would provide a basis for determining how key characteristics of flow affect processes, life stages, and population abundance of FD. A specific set of scenarios could be designed to determine what conditions of low flows lead to high risk for FD. For example, simulations can be run that vary the frequency of occurrence and timing during the year of 5 and 10 days of low flows, with and without delayed spring flows and in combination with some years being drought years.

Systematically Vary Process Rates

This scenario would involve varying the growth, mortality, reproduction, and movement rates of the individual FD within the model under a suite of flows and other environmental conditions. The idea is to create a map of life stage and process sensitivities, which could also be further defined dependent on flow and spatial region. Each M&M measure can then be viewed as affecting certain life stages and processes, at certain times during the year, and in certain spatial areas. How well the M&M measures match up with sensitive life stages, processes, seasons, and areas can guide monitoring to ensure a high likelihood of detecting local responses to management actions. The results can also be used as part of integration to see how well the portfolio (i.e., the mix of M&M measures and minimum flows) covers important life stages and processes. For example, a portfolio that includes three M&M measures that overlap greatly in affecting reproduction of FD via habitat changes, but without at least one measure affecting growth in a critical life stage, can diminish the likelihood of a population response. This set of scenarios can identify redundancies, weaknesses, and gaps in the portfolio of M&M measures and suggest modifications or additions to the measures to increase the probability of redundancy in critical stages and of causing a population response.

Effects of Environmental and Biological Factors

The factors of interest in scenarios do not have to appear in the model to be evaluated. Factors like low dissolved oxygen, sediment removal, algal blooms, gill parasites, and shifts in prey and predator composition can all be examined with the FD model. The M&M measures can be used to determine the magnitude, process, life stage, and location of the likely effects. Most all of the effects of the non-flow-related and perhaps some SAV-related M&M measures would need to be specified outside of the model and then used to change inputs and process representations within the model. For example, if the likely effects of an algal bloom are to reduce food sources, then this can be simulated by reducing growth rates for individuals when they are in the region of the grid where the algal bloom is assumed to occur. The effects of gill parasites could be represented in the model as reduced swimming ability and mortality for larvae and juveniles in historically infected areas. Because some of these changes in factors are done implicitly, the manipulation of the model inputs needs to be done carefully to ensure that the results can be labeled, for example, as the "effects of an algal bloom" and "effects of an M&M measure." Environmental and biological changes can be done singly and in combinations. Generally, the results of this type of exploratory scenario are best viewed as a screening level to provide a rough idea of whether further, more refined, analyses about that factor are warranted.

Vegetative Habitat

Designing scenarios to explore how vegetative habitat affects FD is difficult at this time. The SAV model that would allow for dynamic and spatially explicit responses of SAV to management actions is not yet operational. Rather, the FD model presently uses observed SAV maps and switches them every six months in simulations to match the historical progression of the observed maps. With the present FD model set-up, one could explore vegetation-related M&M measures and other management actions in several ways. One way is to use an implicit approach and keep the observed habitat maps in simulations but adjust growth, mortality, or reproduction of the FD individuals to reflect when they are in the areas where SAV is expected to respond to the management actions. A second way would be to use the existing maps and manipulate them to reflect expected changes based on the management actions; this is challenging to implement, which is why the dynamic SAV model is being developed. However, given proposed changes in goals for SAV acreage (BIO-WEST and Watershed Systems Group, 2016), a first effort to evaluate the impact of changed coverage by native versus non-native SAV species on FD populations could represent a useful application of the model for management purposes. A third approach would be to switch the timing of the existing maps within simulations to determine whether simulated FD population dynamics are sensitive to subregional scale and interannual variability in the observed SAV (habitat) record. One could create specific time series of habitat maps that represent six-month periods of "poor" and "good" habitat maps to ask, for example, how do multiple years of "poor" habitat conditions affect the FD population abundance? Similarly, one could use "good habitat" SAV maps in sequence to roughly represent how restoration of SAV would benefit FD. Finally, the habitat maps could be switched in combination with different flow patterns to quantify any interaction effects between habitat maps and flow.

Spatially explicit models are frequently sensitive to the scale at which state variables and forcings are defined. Beyond the suggested scenarios for forcing vegetation maps to garner insights from the FD model, additional simulations can be designed to evaluate the sensitivity of FD to SAV. For example, are there measureable thresholds of SAV acreage in a given reach that result in dramatic increases or declines in FD abundance? If this is the case using forced maps, how might those insights inform the requirements from the SAV model for a coupled modeling framework that is most effective?

Forced Population Reductions and Density Dependence

Simulations in this scenario would force FD population reductions (simply remove individuals on a day in certain areas) and determine the time period that the population remains below a threshold and the subsequent rate of recovery of the population to a healthier value. The fact that the model has very limited density dependence (see NASEM, 2016) constrains the analysis to short-term predictions. Shorter-term predictions are typically more influenced by the state of the population at the time of stress, whereas long-term predictions are influenced by the density-dependence in the model.

General Suggestions Regarding SAV Model Scenarios

As mentioned previously, the SAV model is not yet far enough along in its development for detailed suggestions regarding scenarios. For example, it is not yet clear if sexual or vegetative reproduction will be successfully represented in the dynamic SAV model, and general information regarding sensitivity analyses that should be used to inform the limits and expectations for model runs are not yet available. However, should the SAV model be successfully launched for these systems, the following general ideas for model applications are offered. As for the FD model, a critical question would appear to be running the model under low flows and for flow protection measures to evaluate the impact on predicted SAV. Further pushing the model to catastrophic scenarios—for example where SAV is only present in refugia—might also reveal some insights regarding recovery following such an event. Clearly, this scenario would require confidence in the model formulations and approaches for simulating reproduction. Although the section above suggested forcing simulated maps of SAV representative of "good" and "bad" years in various virtual time series in the FD model, examining these same questions in a dynamic SAV model would no doubt lead to insights regarding the degree to which the spring and river systems are sensitive to consecutive years of drought. A guiding question behind such a series of simulations might be "how many consecutive years of drought or low flow protection measures can the system withstand?" One of the strengths of the SAV model will be its ability to evaluate M&M measures and help to inform associated adaptive management decisions. Here it would seem valuable to use the model to better understand the degree of long-term maintenance that might be required to eradicate non-native species (how much *Hydrilla* must be removed before the population comes to a steady state at a small enough coverage to be considered controlled in this system?). Are there lessons from the model that can be used to evaluate the timing of planting or non-native vegetation removal that might serve as testable restoration methods that could help optimize the vegetation removal and planting programs? These scenarios are largely predictive in nature, providing output that can be used to evaluate various protective measures or inform improved restoration. However, the EAA is encouraged to explore the diagnostic abilities of this mechanistic model to better understand the environmental forcings that influence vegetation, and to identify future applied research questions that might best serve management goals.

CONCLUSIONS AND RECOMMENDATIONS

Prior to the release of this report, the Committee provided an evaluation of the progress to date on the ecological modeling efforts of the EAA (see Appendix A). Indeed, that short report (NASEM, 2016) covers progress made through mid-2016, including an evaluation of model objectives and usage, configuration, calibration and testing, and submodel coupling, while much of the above text deals with the EAA's response to the Committee's first report (NRC, 2015) evaluating the 2014 year. As stated in

NASEM (2016), the Committee feels that the ecological modeling efforts have made good progress and that scientifically sound frameworks and approaches for the SAV and FD models are in place. For the SAV model, where this report comes in the midst of model development, we send a general message of encouragement. Individual-based models are challenging and complex, in this case several novel solutions are being explored, and continued support of the models will likely lead to very useful products in support of the HCP. The Committee and the EAA model development team both recognize that model development is an iterative process, and so it is expected that the models will continue to reflect new knowledge and understanding with time. The Committee is encouraged by the Applied Research focus on the CSRB for 2016 (see Chapter 5) and looks forward to further assisting the EAA with respect to the Committee's evaluation of the ecological modeling detailed in NASEM (2016). The following conclusions and recommendations refer exclusively to the material in this chapter.

As requested in NRC (2015), **the EAA has now provided a scientifically sound basis for the development of a generalized ecosystem-based conceptual model.** The conceptual diagrams produced to date for the FD and SAV ecological models will help to guide further development of whole-system conceptual models. This collection of conceptual models will provide a communication tool for the HCP, will aid in coordination of the diverse expertise found across EAA's multiple advisory committees and contractors, and will serve an important function, along with the predictive ecological models, to evaluate the appropriateness and efficacy of the M&M measures.

The EAA is making progress on addressing the sampling deficiencies that may limit the ability to estimate the distribution and abundance of CSRB populations. The focus on the CSRB in the 2016 and 2017 Applied Research Program is a substantial effort for addressing the limited knowledge about the distribution and life history features that will be important for understanding how the CSRB responds to environmental variation, including changes in flow and responses during drought conditions. If the CSRB is to remain an indicator taxon for other listed invertebrate and vertebrate species, these gaps in life history and distribution will need to be addressed. Alternatively, the EAA should begin to develop monitoring plans for the other listed species.

The continued development of the FD and SAV models will result in models that can address a wide variety of questions about the effectiveness of flow protection and other M&M measures. The models offer a very powerful tool for combining multiple effects across life stages and space into ecologically relevant end points. Reaping the benefits of the ecological models will likely involve continuing, in some manner, the ecological modeling program beyond the originally anticipated time frame.

Armed with a fully capable FD model, the scenarios analyzed should be designed and documented according to the concepts in this chapter. These include careful designing of the scenarios and use of terminology to ensure transparency, confirming scenarios are within the domain of applicability, associating uncertainty with model predictions, and properly interpreting predictions and providing model-based mechanistic explanations for model responses.

Seven scenarios are described for the fountain darter model, which can be either diagnostic based (e.g., varying process rates) or evaluative (e.g., running the bottom-up package). The scenarios offered demonstrate how the model can be used to examine how extreme flows, process rates, environmental factors, SAV habitat, and episodic population reductions affect FD population dynamics. These results can then be merged with the expected effects of M&M measures to identify the robustness and redundancies of the entire suite of actions.

Only general guidance is given on possible scenarios for the SAV model, as it is not appropriate to provide detailed advice at this stage of model development. Nonetheless, given the recently proposed adaptive management actions related to changing SAV species coverage goals in the HCP, it would be timely to evaluate the longer-term impact of these decisions on the stability of the SAV populations. The prospect of having such a valuable quantitative tool to better understand the effects of M&M measures and predict future states will hopefully motivate those involved to continue developing the SAV model.

REFERENCES

Addison, P. F., L. Rumpff, S. S. Bau, J. M. Carey, Y. E. Chee, F. C. Jarrad, M. F. McBride, and M. A. Burgman. 2013. Practical solutions for making models indispensable in conservation decision-making. Diversity and Distributions 19:490-502.

Balaguer, L., A. Escudero, J. F. Martín-Duque, I. Mola, and J. Aronson. 2014. The historical reference in restoration ecology: re-defining a cornerstone concept. Biological Conservation 176:12-20.

BIO-WEST. 2015. Predictive ecological model for the Comal and San Marcos ecosystems project. Edwards Aquifer Habitat Conservation Plan. Interim Report. Contract No. 13-637-HCP.

BIO-WEST, Inc. and Watershed Systems Group, Inc. 2016. Submerged Aquatic Vegetation Analysis and Recommendations. Edwards Aquifer Habitat Conservation Plan Contract No. 15-7-HCP June, 2016.

Bjørkvoll, E., V. Grøtan, S. Aanes, B. E. Sæther, S. Engen, and R. Aanes. 2012. Stochastic population dynamics and life-history variation in marine fish species. The American Naturalist 180:372–387.

Bradfield, R., G. Wright, G. Burt, G. Cairns, and K. Van Der Heijden., 2005. The origins and evolution of scenario techniques in long range business planning. Futures 37:795-812.

Duarte, C. M., A. Borja, J. Carstensen, M. Elliott, D. Krause-Jensen, and N. Marbà. 2015. Paradigms in the recovery of estuarine and coastal ecosystems. Estuaries and Coasts 38(4):1202-1212.

EAA. 2015. National Academy of Sciences—Review of the Edwards Aquifer Habitat Conservation Plan. Report 1 Implementation Plan. Edwards Aquifer Authority August 20, 2015.

EARIP. 2012. Habitat Conservation Plan. Edwards Aquifer Recovery Implementation Program.

Ferson, S., and L. R. Ginzburg. 1996. Different methods are needed to propagate ignorance and variability. Reliability Engineering & System Safety 54:133–144.

Guisan, A., R. Tingley, J. B. Baumgartner, I. Naujokaitis-Lewis, P. R. Sutcliffe, A. I. Tulloch, T. J. Regan, L. Brotons, E. McDonald-Madden, C. Mantyka-Pringle, and T. G. Martin. 2013. Predicting species distributions for conservation decisions. Ecology Letters 16(12):1424-1435.

Hamilton, S. H., C. A. Pollino, and A. J. Jakeman. 2015. Habitat suitability modelling of rare species using Bayesian networks: Model evaluation under limited data. Ecological Modelling 299:64-78.

Hardy, T. B., K. Kollaus, and K. Tower. 2010. Evaluation of the Proposed Edwards Aquifer Recovery Implementation Program Drought of Record Minimum Flow Regimes in the Comal and San Marcos River Systems. River Systems Institute, Texas State University.

Higgs, E., D. A. Falk, A. Guerrini, M. Hall, J. Harris, R. J. Hobbs, S. T. Jackson, J. M. Rhemtulla, and W. Throop. 2014. The changing role of history in restoration ecology. Frontiers in Ecology and the Environment 12:499-506.

Lempert, R. 2013. Scenarios that illuminate vulnerabilities and robust responses. Climatic Change 117:627-646.

Letcher, B. H., and J. A. Rice. 1997. Prey patchiness and larval fish growth and survival: Inferences from an individual-based model. Ecological Modelling 95:29–43.

Link, J. S., T. F. Ihde, C. J. Harvey, S. K. Gaichas, J. C. Field, J. K. T. Brodziak, H. M. Townsend, and R. M. Peterman. 2012. Dealing with uncertainty in ecosystem models: The paradox of use for living marine resource management. Progress in Oceanography 102:102–114.

Magnusson, A., A. E. Punt, and R. Hilborn. 2013. Measuring uncertainty in fisheries stock assessment: The delta method, bootstrap, and MCMC. Fish and Fisheries 14:325–342.

Merow, C., M. J. Smith, T. C. Edwards, A. Guisan, S. M. McMahon, S. Normand, W. Thuiller, R. O. Wüest, N. E. Zimmermann, and J. Elith. 2014. What do we gain from simplicity versus complexity in species distribution models? Ecography 37(12):1267-1281.

NASEM (The National Academies of Science, Engineering, and Medicine). 2016. Evaluation of the Predictive Ecological Model for the Edwards Aquifer Habitat Conservation Plan: An Interim Report as Part of Phase 2. Washington, DC: The National Academies Press.

NRC (National Research Council). 2015. Review of the Edwards Aquifer Habitat Conservation Plan: Report 1. Washington, DC: The National Academies Press.

Papworth, S. K., J. Rist, L. Coad, and E. J. Milner-Gulland. 2009. Evidence for shifting baseline syndrome in conservation. Conservation Letters 2(2):93-100.

Parson, E. A., V. Burkett, K. Fischer-Vanden, D. Keith, L. O. Mearns, H. Pitcher, C. Rosenzweig, and M. Webster. 2007. Global-change scenarios: Their development and use, synthesis and assessment product 2.1b. U.S. Climate Change Science Program.

Pauly, D. 1995. Anecdotes and the shifting baseline syndrome of fisheries. Trends in Ecology and Evolution 10(10):430.

Peck, S. L. 2004. Simulation as experiment: a philosophical reassessment for biological modeling. Trends in Ecology and Evolution 19(10):530-534.

Rose, K. A., S. Sable, D. L. DeAngelis, S. Yurek, J. C. Trexler, W. Graf, and D. J. Reed. 2015. Proposed best modeling practices for assessing the effects of ecosystem restoration on fish. Ecological Modelling 300:12-29.

Saltelli, A., and P. Annoni. 2010. How to avoid a perfunctory sensitivity analysis. Environmental Modelling & Software 25:1508–1517.

Saltelli, A., S. Tarantola, F. Campolongo, and M. Ratto. 2004. Sensitivity Analysis in Practice: A Guide to Assessing Scientific Models. West Sussex, England: John Wiley and Sons.

Stow, C. A., J. Jolliff, D. J. McGillicuddy, S. C. Doney, J. Allen, M. A. Friedrichs, K. A. Rose, and P. Wallhead. 2009. Skill assessment for coupled biological/physical models of marine systems. Journal of Marine Systems 76:4–15.

Zara Environmental. 2015. Comal Springs Riffle Beetle Occupancy Modeling and Population Estimate within the Comal Springs System, New Braunfels, Texas. Prepared by Zara Environmental LLC and submitted on 23 March 2015.

4

Biological and Water Quality Monitoring

The Committee made several comments and recommendations associated with the design, purpose, integration, and adequacy of the water quality and biological monitoring programs in its first report (NRC, 2015). In particular the Committee raised concerns about the apparent lack of integration between the water quality and biological monitoring programs, the difficulty of making system-wide estimates of target species population densities and trends given the reliance on non-randomized sampling of selected index reaches, the inability to assess whether changes in nutrient status are leading to changes in the frequency and magnitude of algal blooms because of insufficient detection limits of phosphorous and nitrogen, and the inability to determine population densities and spatial distribution of the invertebrate target species such as the Comal Springs riffle beetle.

In response to the Committee's recommendations, the Edwards Aquifer Authority (EAA) established two working groups to assess the water quality and biological monitoring programs, respectively, and make necessary modifications, and they added a Ph.D. level scientist (Dr. Chad Furl) to its staff to assist with these efforts. This evaluation of the two monitoring programs provided the EAA with an opportunity to integrate more closely the water quality and biological monitoring programs to provide efficient and seamless measurement of variables important to inform the modeling efforts and ensure that species of interest maintain adequate population levels. These working groups met throughout the spring of 2016 and issued a joint report (EAHCP, 2016) for consideration by the EAA Implementation Committee. As shown in Tables 4-1 and 4-2, the working groups comprised representatives from various stakeholder groups and included members of the Sci-

TABLE 4-1 2016 Water Quality Monitoring Program Work Group

Name	Organization
Ken Diehl	San Antonio Water System
Melani Howard	City of San Marcos/Texas State University
Charlie Kreitler	Science Committee
Steve Raabe	Stakeholder Committee/San Antonio River Authority
Ben Schwartz	Texas State University
Mike Urrutia	Guadalupe-Blanco River Authority

TABLE 4-2 2016 Biological Monitoring Program Work Group

Name	Organization
Tyson Broad	Texas Tech University
Jacquelyn Duke	Science Committee/Baylor University
Mark Enders	City of New Braunfels
Rick Illgner	Edwards Aquifer Authority
Doyle Mosier	Science Committee

ence Committee. While there was no overlap in membership between the two working groups, Steven Raabe from the San Antonio River Authority was appointed as joint chair of both the water quality and biomonitoring working groups, presumably in an effort to coordinate recommendations between the two working groups. It appears that the ecological modeling team was not represented in these working groups. This is unfortunate because inclusion of one or more members of the modeling team would have allowed for better integration between the modeling and monitoring efforts, which is important for ensuring that the data collected by the monitoring programs are directly useful in the model calibration and validation efforts. The two working groups have now disbanded, having completed their tasks. The EAA should consider forming a standing working group on monitoring that would meet as needed to provide advice and outside perspective on the EAA's monitoring program.

The joint report of the two working groups (EAHCP, 2016) presents a number of modifications to the existing water quality and biomonitoring programs (see Tables 4-3 and 4-4). The revised monitoring program eliminates monitoring of a large list of contaminants that have not been found to occur in detectable concentrations in the spring systems, and adds sampling of fish tissue for particular contaminants, one additional sonde measure-

TABLE 4-3 Final Recommendations for the Water Quality Monitoring Program

Sampling Method	Final Recommendations	Justification
Surface water (base flow)	Remove from program	• Sampled by Clean Rivers Program • No significant detects • BioMP collects field and nutrients water quality at low and high flow
Sediment	Biennially in even years	• Data will change little throughout the year • Biological monitoring data do not suggest impact to Covered Species • Provides information on water quality trends in toxic parameters
Real-time monitoring	Add one monitoring station per system	• Valuable source of continuous information that is ecologically relevant • Field parameters collected every 15 minutes: dissolved oxygen, conductivity, turbidity, temperature, pH
Stormwater	Reduce to one sampling event each year; Test only for IPMP chemicals in odd years, test full suite in even years as currently done, add two samples to the rising limb of the hydrograph for a total of 5 samples/location; priority given to locations at tributary outflows	• Turnover rate, dilution • Lack of significant detects
Passive diffusion sampling (PDS)	Add PPCP membrane only at bottom of channel	• PDS provides a sensitive index for contamination in the spring systems
Groundwater (well)	Remove from program	• Purpose is to detect movement of bad water line • Already sampled by EAA
Tissue sampling	Add to program, one sample in odd years	• Represents direct link to Covered Species • Parameters and species to be established (work with experts) • Provides new information and data • Species to be sampled will be determined in consultation with experts

BioMP = biological monitoring program
EAA = Edwards Aquifer Authority
IPMP = Integrated Pest Management Plan
PPCP = pharmaceutical and personal care products
SOURCE: EAHCP (2016).

TABLE 4-4 Final Recommendations for the Biological Monitoring Program

Sampling Methods	Final Recommendations	Justification
Fixed station photography	No modification	Valuable historical baseline
Aquatic vegetation mapping, including Texas wild rice	No modification	Valuable baseline, trend and compliance information
Fountain darter sampling	No modification	Valuable indices to fish population health
Fish community sampling	No modification	Provides macro information pertinent to Covered Species
Invertebrate sampling – Covered Species	No modification	Provides macro information pertinent to Covered Species
Macroinvertebrate food source monitoring	Substitute RBAs • Use TCEQ/TPWD RBA Option 1 Protocol for macroinvertebrate community health without variables. • Frequency and locations: Samples the five (5) Reaches in Comal system; four (4) reaches in San Marcos system. One (1) composite sample per reach. Thus, total of nine (9) samples for both systems per Comprehensive and Critical Period Event. • Sampling details: The result is only one sample per reach. • Logistics: To be conducted at the same time as fixed drop-net sampling for fountain darters. • Procedural details: Collect and identify (to lowest practical taxonomic level) first one hundred (100) macroinvertebrates.	Cost: More economical option Programmatic: More consistent with requirements of biological monitoring program.
Salamander visual observations	No modification	Necessary to monitor population health
Comal Springs discharge measurement	No modification	Important environmental measure
Flow partitioning within Landa Lake	Remove from program	To be done through EAA

TABLE 4-4 Continued

Sampling Methods	Final Recommendations	Justification
Water Quality grab sampling	Continue to collect but modify method detection limit (MDL) for SRP from 50 µg/L to 5 µg/L	Continue—important accompaniment to biological information
Critical period (high and low-flow events)	No modification	Important index during critical periods

EAA = Edwards Aquifer Authority
RBA = rapid bioassessment
SRP = soluble reactive phosphorus
TCEQ = Texas Commission on Environmental Quality
TPWD = Texas Parks and Wildlife Department
SOURCE: EAHCP (2016).

ment station in each spring system, and rapid bioassessment protocols. The resulting water quality and biomonitoring programs are better integrated, more targeted to the species of concern, more efficient, and provide more standardized monitoring of the overall health and quality of the aquatic ecosystems. A detailed discussion of the progress made on monitoring the CSRB is provided in Chapter 5.

Nutrients. As a result of the Committee's recommendations in NRC (2015) and the deliberations of the monitoring working groups, the EAA has made a number of modifications to the monitoring of nutrients (nitrogen and phosphorus). Of particular note is the lowering of the detection limit for phosphorous. The detection limit for soluble reactive phosphorus will be lowered to 3-5 µg/L from the current detection limit of 50 µg/L. The lower detection limit for soluble reactive phosphorus will enhance the ability to detect increasing or decreasing trends of what is likely the limiting nutrient in the system and help provide an early warning of eutrophication, which can lead to depleted levels of dissolved oxygen.

The detection limit for nitrogen species will remain at 50 µg/L for nitrate and 100 µg/L for ammonia. The working group recommended that the detection limits not be changed for nitrogen species after examining data collected thus far and finding that almost all values were well above the detection limits. The EAA will also partner with the Clean Rivers Program, which also does routine monitoring of nutrients in the spring and river systems. In particular, it appears that the EAA will rely on Clean Rivers Program data for nitrogen and for total phosphorus, while continuing to collect data for soluble reactive phosphorus in house. It appears that the Clean Rivers Program uses adequate methods and quality assurance/

quality control protocols. It was not clear, however, whether the Clean Rivers Program samples at the same locations and frequency as the EAA. It is important that these sampling efforts are co-located in space and time so that the data can be used to assess nutrient effects on the spring and river systems. In addition, to enable future interpretation of nutrient monitoring data, it is important that the many analyses be performed on the same water sample.

PAHs. Prior monitoring efforts have shown that the level of contaminants is generally low in the spring and river systems, with one important exception. Recently, the concentration of polycyclic aromatic hydrocarbons (PAHs) was detected at levels as high as 62 mg/kg in a location in the San Marcos River (Blanton and Associates, 2016). This is well above a probable effects concentration of 22.8 mg/kg designed to protect biota (MacDonald et al., 2000). PAHs exposure leads to a narcosis reaction in invertebrates that can result in adverse effects from mild disruption of cell membranes to mortality (Burgess, 2007). If found to be widespread throughout the river system (which is not suggested by the current data) elevated PAHs could lead to significant risks to the listed species that are not currently being addressed by the sediment removal efforts or other monitoring and habitat protection measures. Efforts to identify the significance of these elevated PAH sediments should be undertaken. The frequency and extent of high concentrations of PAHs should be established by more extensive sampling in areas where elevated levels have been identified.

A source of PAHs in urban areas without significant point sources is coal tar-sealed parking lots. Indeed, coal tar sealants may constitute the vast majority of PAHs in adjacent sediments (MacDonald et al., 2000). The EAA banned coal tar as a parking lot sealant over the aquifer's recharge zone in Hays and Comal Counties in 2012. San Marcos passed a coal tar sealant ban in May 2016, while San Antonio passed a coal tar ban in July 2016. New Braunfels does not have a coal tar ban. If the parking lot sealant is the source of the PAHs observed in the San Marcos River, these bans will lead to reduced sediment PAH concentrations over time.

The actual risks to listed species as a result of exposure to elevated PAH levels is unknown. This is because substantial quantities of PAHs, particularly from coal tars, may be tied up in largely non-bioavailable forms (Cornelissen et al., 2005), such that their presence in sediments does not necessarily suggest that elevated risks are present. If it is not possible to substantially reduce PAH concentrations through sediment removal and source control, evaluation of bioavailability of the PAHs in the sediment should be considered.

Performance Monitoring of Minimization and Mitigation Measures. One addition to the monitoring program is the requirement that groups involved in riparian habitat improvement institute monitoring to assess the effectiveness of the improvements (EAHCP, 2016). This is a step in the right direction, and it directly addresses a recommendation of NRC (2015) to performance monitor the minimization and mitigation (M&M) measures. Ideally, all M&M measures that are implemented as part of the Habitat Conservation Plan (HCP) should be integrated into one conceptually unified monitoring program. These M&M measures are often multi-year in scope, such that it may take additional years of monitoring to evaluate the success of the measures. It would be best if the performance monitoring of M&M measures could be integrated into the existing water quality and biological monitoring programs. This vision conceptualizes monitoring as one multi-faceted program that collectively addresses information needs associated with water quality, biology, modeling, and M&M measures.

For example, the EAA will add one continuously recording water quality sonde to each river system. Data from these sondes will provide valuable information on dissolved oxygen and other parameters. In addition, the EAA should consider deploying the miniDOT dissolved oxygen sensors used in the Landa Lake dissolved oxygen study as part of the routine monitoring program. These data, in conjunction with the multiparameter water quality sonde, will provide important, highly resolved spatial and temporal data on dissolved oxygen in Landa Lake. This is an example of how selected measurements first made during Applied Research projects can be integrated into the monitoring program.

CONCLUSIONS AND RECOMMENDATIONS

The main goal for the water quality and biological monitoring programs should be to develop a single, integrated program that provides the basic information needed to assess compliance with the HCP. The monitoring programs are designed to provide long-term data that will allow the EAA and others to assess trends in water quality and biology. The following specific recommendations suggest a few steps in this direction.

The monitoring program should include the measurements needed to monitor the performance of the broad suite of minimization and mitigation measures. Relying on the individual Applied Research projects or M&M activities to provide these data is unrealistic as these projects and measures are not designed nor funded over the long term, even though it may well take multiple years for the effects of these projects to be realized.

The monitoring program should include the long-term data required to test and inform continuous refinements of the ecological model. The ecological model will need to be continuously assessed and refined, and long-term data collected by the monitoring program will be critical to this effort. It is important that the modeling team be involved in the design of the monitoring program to ensure that the variables being measured are the ones that are most important for model assessment.

The frequency and extent of high concentrations of PAHs should be established by more extensive sampling in areas where elevated levels have been identified. If it is not possible to substantially reduce PAH concentrations through sediment removal and source control, evaluation of bioavailability of the PAHs in the sediment should be considered.

REFERENCES

Blanton and Associates. 2016. Edwards Aquifer Habitat Conservation Plan 2015 Annual Report. Submitted to the U.S. Fish and Wildlife Service March 22, 2016.

Burgess, R. M. 2007. Evaluating Ecological Risk to Invertebrate Receptors from PAHs in Sediments at Hazardous Waste Sites. U.S. Environmental Protection Agency. Office of Research and Development. National Health and Environmental Effects Research Laboratory. Atlantic Ecology Division. EPA/600/R-06/162.

Cornelissen, G., Ö. Gustafsson, T. D. Bucheli, M. T. O. Jonker, A. A. Koelmans, and P. C. M. van Noort. 2005. Extensive sorption of organic compounds to black carbon, coal, and kerogen in sediments and soils: Mechanisms and consequences for distribution, bioaccumulation, and biodegradation. Environmental Science & Technology 39(18):6881-6895.

EAHCP. 2016. Report of the 2016 Expanded Water Quality Monitoring Program Work Group and Report of the 2016 Biological Monitoring Program Work Group.

MacDonald, D. D., C. G. Ingersoll, and T. A. Berger. 2000. Development and evaluation of consensus-based sediment quality guidelines for freshwater ecosystems. Archives of Environmental Contamination and Toxicology 39(1):20-31.

NRC (National Research Council). 2015. Review of the Edwards Aquifer Habitat Conservation Plan: Report 1. Washington, DC: The National Academies Press.

5

Applied Research Program

EAA RESPONSE TO COMMITTEE'S FIRST REPORT

In NRC (2015) a number of broad recommendations and conclusions were made about the Applied Research Program that covered three general areas: improving the process used to solicit, review, and manage the Applied Research Program; adopting and implementing a data management system; and increasing understanding of the Comal Springs riffle beetle. The Committee was pleased to learn that, in general, the Edwards Aquifer Authority (EAA) responded by implementing most of its recommendations in these areas, as discussed below.

Improvements to the Applied Research Program Process

The Committee recommended that the process of identifying, soliciting, reviewing, and selection of projects be more transparent by soliciting additional proposals from new proposers and asking for more input from the Science Committee on the key elements to be included in the requests for proposals as well as on the technical merits of resulting proposals. The EAA responded by appointing an Applied Research Working Group and modifying their procedures for soliciting, reviewing, and awarding research projects (EAA, 2015). These new procedures were in place for the selection of the 2016 Applied Research projects.

In addition, the EAA also took steps to try to increase the number of outside experts submitting proposals to the Applied Research Program by broadly advertising the research solicitations. In addition, the EAA used literature and web searches to identify scientists whose interests and expertise aligned with the subject areas of the 2016 request for proposals and made sure these potential applicants were aware of the research opportunities. It is unclear whether these efforts led to additional involvement of outside experts in the Applied Research Program. The EAA is encouraged to persist in these efforts to attract interested experts who are currently not involved in the various Habitat Conservation Plan (HCP) efforts by looking for ways to remove conditions that might overly restrict the pool of potential applicants. For example, the Committee noted that the time period from release of the request for proposals to the deadline for proposal receipt for the 2017 program was relatively short, making it difficult for experts not familiar with the EAA to respond to the requests for proposals.

Finally, the Committee recommended that a process be implemented to allow Applied Research projects to have a duration of more than one year if needed to meet the goals of the study. The EAA and the Applied Research Working Group are working to implement this recommendation, with one current project on the Comal Springs riffle beetle (CSRB) life history currently having a two-year duration.

Implement a Database Management System

Through the Applied Research Program and the monitoring program the EAA has collected a large of amount of data covering a wide variety of physical, chemical, biological, and hydrologic variables. In NRC (2015) the Committee recommended that these data be organized, stored, and curated in a database management system to ensure that all data are available in a well-documented manner to both internal and external parties. The EAA has responded positively to this recommendation. They hired a new database manager and are now using an off-the-shelf database management system called Aquatic Informatics–Aquarius Samples. The data management staff is currently working to populate this system with the accumulated data. The Committee applauds this action as it will make data analyses and modeling calibration and validation efforts more efficient and streamlined going into the future. This system will also serve as a long-term data repository and archival mechanism. The Committee continues to encourage the EAA to use the data management system being implemented to allow greater data discovery and access by the outside scientific community and the public. Access to the core monitoring data can lead to greater involvement of stakeholder groups, encourage participation by a broader group of experts, and result in a better overall program.

Recommendations for Additional CSRB Studies

The EAA has made considerable effort to implement the Committee recommendations for additional Applied Research studies regarding the biology of the CSRB, with most/all of the 2016 projects focused on understanding the CSRB. The three 2016 proposals include studies on the CSRB functional group classification, tolerances to high temperatures and low oxygen conditions, and life history:

1. *Evaluation of the Trophic Status and Functional Feeding Group Status of the Comal Springs Riffle Beetle* prepared by Weston Nowlin and Dittmar Hahn of Texas State University (EAHCP Proposal # 148-15-HCP).
2. *Evaluation of Long-Term Elevated Temperature and Low Dissolved Oxygen Tolerances of the Comal Springs Riffle Beetle* prepared by Weston Nowlin and Benjamin Schwartz of Texas State University (EAHCP Proposal # 146-15-HCP).
3. *Evaluation of the Life History of the Comal Springs Riffle Beetle* prepared by BIO-WEST Project Team (EAHCP Proposal # 147-15-HCP).

The first project, which would evaluate the trophic and functional feeding group status of the CSRB, may contribute to a better understanding of the habitat breadth of the species. The proposal argues that the CSRB populations are largely restricted to areas adjacent to spring outflows, areas of the benthic habitat that also collect substantial amounts of coarse organic matter in the form of leaves and small branches, and that these food resources also serve as habitat. If this is indeed the case, the proposed project has the potential to reveal an alternative sampling approach for the CSRB that could be a reasonable reflection of population densities. Using stable isotopes the proposed research would identify the primary food resources of the CSRB, and if linked to the primary habitat of larvae and adults would reveal a habitat that could be sampled quantitatively. This proposed research project is one of the first to take a novel approach to identifying the food resources and habitat of the beetle with potentially transformative results that could lead to quantitative population estimates important for monitoring and ecological models.

The second project listed above was not directly related to the recommendations put forth by the Committee in NRC (2015), namely the project evaluating temperature and oxygen requirements of the CSRB. Understanding the tolerance values of the CSRB has merit for future estimates of how the species will respond to changes in abiotic conditions; however, without an objective and quantitative method for estimating the population size, models that incorporate these variables will not provide additional information for better population predictions. The proposed project would provide new data on how individual beetles respond to increased water temperature and associated decreased dissolved oxygen in a series of laboratory studies. The project would also evaluate the tolerances of surrogate species for the CSRB. There are three potential concerns for this study. As described in NRC (2015), there are potentially significant biological differences among Elmidae species, so that the use of surrogate species may not adequately represent the CSRB. Similarly, the results from laboratory studies do not always represent how a species will respond in the natural environment, an issue likely important for the CSRB given that it is considered a subterranean organism that comes to the benthic surfaces for some feeding. The last concern is that the laboratory facilities that were proposed have suffered large mortalities of the CSRB adults, with water quality changes the suspected cause for these deaths. At the time of this writing, the cause has not been definitively identified and the issue remains unresolved.

The third proposed research project is on CSRB life history, which is an important effort for understanding how the populations of the CSRB change naturally as part of the life cycle and also how they may respond to changing environmental conditions. Also, understanding the life history and life cycle characteristics will allow scientists to evaluate the results from previous and ongoing efforts to measure, and eventually predict, the population abundances of the CSRB. It should be noted that the proposed project on CSRB life history characteristics will be conducted at the same laboratory facility as that of the temperature and oxygen study, and may suffer from the same potential issues as described above.

A major recommendation for the Applied Research Program from NRC (2015) was to better (1) quantify the CSRB population densities and/or (2) calibrate the cotton-lure method of sampling so that it could potentially be an efficient and reliable way to estimate populations. The inability to calibrate the cotton-lure method of sampling with any real densities of the CSRB in the system is a considerable weakness, making the representativeness of this sampling approach for estimating population densities unknown and making monitoring for CSRB population estimates difficult if not impossible to achieve. If the species population cannot be estimated with some degree of certainty and account for responses to flow variability, its status as an indicator species for other endangered or threatened species in the ecosystem should be re-addressed.

Recognizing the weaknesses of the CSRB sampling, the Applied Research Work Group created a fourth 2016 Applied Research study on CSRB sampling (EAHCP ARWG, 2015), although no proposal was written. To fulfill this project, a regional workgroup was created to establish a Standard Operation Procedure (SOP) for how to deploy, retrieve, and score cotton lures when collecting CSRBs. The main goal was to standardize data collected by all entities, for entry into a newly created database. This work group, called the cotton-lure SOP workgroup, was attended by all in the region that collect CSRBs, including HCP contractors and staff, and it is not a work group of the HCP. The main deliverable was an SOP and field data sheet that all entities agreed to use.

The goals of developing a standard operating procedure for using the cotton lure sampling approach for the CSRB and the creation of an associated long-term database are to be commended. The plan consists of construction and materials used for the cotton lures, the arrangement of the cotton lure within a wire cage that is used for deployment, and data collection protocols for the deployment, monitoring, and collection. Although the procedure is complete and takes into account most of the information important for a long-term database, including a standardized data sheet, the following concerns are noted. First, **there should be a method to provide standardized data that accounts for the amount of time that the cotton lure has been deployed.** For example, some lures will be placed for the recommended four weeks, but it is possible that others will be placed for other time periods (e.g., three weeks). One hundred CSRB specimens captured over four weeks is not equivalent to 100 specimens collected in only three weeks. By stating data as the number of CSRB specimens per day or week, the data derived from this sampling approach would be better standardized. Second, the findings from the 2015 CSRB Connectivity Study (see subsequent section) should be utilized, which suggest that terrestrially derived leaf and wood organic matter is important to the CSRB populations. Thus, **it would be useful to identify the type, and estimate the relative amount, of organic matter near or at the cotton lure placement locations.** The SOP data collection sheet already requires that the presence of different organic matter types (e.g., wood debris, leaves, roots, macrophytes) be recorded. This could be modified to include an estimated surface area covered by each organic matter type.

RECENTLY COMPLETED APPLIED RESEARCH

For completeness, Tables 5-1, 5-2, 5-3, and 5-4 list all studies that have been part of the Applied Research Program or have been special studies critical to the ecological model or the implementation of minimization and mitigation measures. Those studies that were completed in late 2014/2015 are discussed in greater detail below. Discussion of the 2013 and 2014 studies can be found in NRC (2015).

TABLE 5-1 Fountain Darter Applied Research Projects

Study Title	Year	Objective
1. Fountain Darter Food Source Study to Determine the Critical Thermal Maximum of *Hyalella azteca*	2013	To determine the critical thermal maximum of *Hyalella azteca*, a supposed fountain darter food source. Final report completed (BIO-WEST and Baylor University, 2013).
2. Effects of Vegetation Decay and Water Quality Deterioration on Fountain Darter Movement	2014	To describe fountain darter movement as a function of water quality and vegetation decay using fluorescent tags. Final report completed (BIO-WEST, 2014a).
3. Effects of Low-Flow on Fountain Darter Fecundity	2014	To determine if changes in physical habitats, especially low-growing and dense vegetation, will reduce the reproductive readiness and success of the fountain darter. Final report completed (Texas State University and BIO-WEST, 2014a).
4. Effects of Predation on Fountain Darter Population Size at Various Flow Rates	2014	To determine if flow conditions may cause different relationships between predator and prey and habitat utilization. Final report completed (Texas State University and BIO-WEST, 2014b).

TABLE 5-2 Submersed Aquatic Vegetation (SAV) and Texas Wild Rice Applied Research Projects

Study Title	Year	Objective
1. Field vs. Laboratory Study—Comparison of the Responses of Three SAV	2013	Preliminary study to compare aquatic vegetation (*Ludwigia, Cabomba,* and *Sagittaria*) growth over time when conducted simultaneously in laboratory and in-situ experiments held at similar flow and water quality conditions. Final report available (BIO-WEST and Baylor University, 2013).
2. Vegetation Tolerance Studies A and B	2013	To evaluate the effects of elevated water temperatures in combination with low CO_2 and minimal flow on *Ludwigia, Cabomba, Vallisneria,* and *Riccia* in the lab and in ponds. Final report available (BIO-WEST and Baylor University, 2013).
3. pH Drift Study—Effects of HCO_3^- utilization by select SAV	2013	To determine which of the major SAV species of the Comal River are capable of utilizing HCO_3^- as a carbon source for photosynthesis. Final report available (BIO-WEST and Baylor University, 2013).

TABLE 5-2 Continued

Study Title	Year	Objective
4. Converting SAV Biomass to Percent Areal Cover	2014	To develop an empirical relationship between vegetation percent cover and biomass for use in the ecological model. This will provide a realistic way to convert percent cover maps to levels of biomass present within the system. Final report available (Doyle et al., 2014).
5. *Ludwigia* Interference Plant Competition Study	2015	To evaluate *Ludwigia repens* growth competition and interference by *Hygrophila* sp. and *Hydrilla* sp. To better understand dispersal of *Ludwigia* and refine biological objectives. Final report completed (Center for Reservoir and Aquatic and Systems Research and BIO-WEST, 2015).

TABLE 5-3 Comal Springs Riffle Beetle Applied Research Projects

Study Title	Year	Objective
1. Extended Low-Flow Period Effects on Comal Springs Riffle Beetle	2014	To study CSRB survivorship inside of the springs during periods of low flow and flow cessation, including associated physical (i.e., temperature) and chemical (i.e., DO, pH, conductivity) changes. They designed "aquaria" that allow replicate samples and manipulation of flows to simulate up-welling, middle-welling, and top-welling. Final report completed (BIO-WEST, 2014b).
2. Determination of Limitations of Comal Springs Riffle Beetle Plastron Use during Low Flow	2014	Adult riffle beetles have fine hairs (plastron) that trap air next to their body, acting as a gill to breath underwater. Plastrons require clean, cool water to function. Determination of the limitations of the plastron to reduced dissolved oxygen levels and elevated temperatures would be useful in habitat management and modeling for the conservation of the CSRB. Final report completed (Gibson et al., 2013).
3. Estimate Comal Springs Riffle Beetle Population in Comal Springs/Landa Lake	2014	Sample a random distribution of previously sampled and unsampled springs for CSRB within Comal Springs/Landa Lake to estimate the CSRB population. Final report completed (Zara Environmental, 2015).

continued

TABLE 5-3 Continued

Study Title	Year	Objective
4. Comal Springs Riffle Beetle Habitat Connectivity	2015	Evaluate the importance of the surface, riparian, and submerged food sources to the ecology of the CSRB at the springs. Final report completed (BIO-WEST and Texas State, 2015).
5. Evaluation of the Long-Term, Elevated Temperature and Low Dissolved Oxygen Tolerances of the Comal Springs Riffle Beetle	2016	Examine the individual and combined roles of relatively long term increases in temperatures and declines in DO concentrations on CSRB and potentially several other riffle beetle species in an experimental laboratory-based setting. Proposal available (Nowlin and Schwartz, 2015).
6. Evaluation of the Life History of the Comal Springs Riffle Beetle	2016	Things to be studied in laboratory environment include eggs, mating, larvae, larval completion, pupation, adult lifespan, and fecundity. Proposal available (Bio-WEST, 2016).
7. Evaluation of the Trophic Status and Functional Feeding Group Status of the Comal Springs Riffle Beetle	2016	Utilize a stable isotope approach to determine the feeding ecology of the CSRB and other invertebrates found in the upper Comal system. Also, characterize the microbial communities associated with biofilms in the Comal and compare them to the communities found within the guts of CSRB and the biofilms found in different microhabitats within Comal Springs. Proposal available (Nowlin and Hahn, 2015).
8. CSRB Quantitative Sampling Techniques	2016	Determine efficacy of different sampling techniques. No proposal available, but a standard operation procedure was created that satisfies as the deliverable for this project.

DO = dissolved oxygen

TABLE 5-4 Other Applied Research Projects or Special Studies

Study Title	Year	Objective
1. Algae Dynamics and Dissolved Oxygen Depletion Study	2015	To better understand the cause and effects of excessive algal blooms on bryophytes in the Upper Spring Run and Landa Lake sections of the Comal River. Final report completed (BIO-WEST, Center for Reservoir and Aquatic Systems Research, and Aqua Strategies, 2015).
2. Preliminary Tests of an Aeration System in Landa Lake	2015	The final report *Supplemental Dissolved Oxygen Evaluation in Landa Lake* (BIO-WEST, 2015).

TABLE 5-4 Continued

Study Title	Year	Objective
3. Suspended Sediment impacts on TWR (and Other SAV) and Macroinvertebrates	2015	To evaluate the timing and duration of suspended sediments in the San Marcos River, to evaluate suspended sediment impact on aquatic plant communities and on the aquatic macroinvertebrate community, and to produce information that will be useful for any eventual TWR model. Final report completed (Texas State University, 2016).
4. Database Creation and Management	2016	Database creation and management, including compiling and formatting data, creating standard data templates, and normalizing data for all EAHCP applied research conducted to date. They are using an off-the-shelf product: Aquatic Infomatics – Aquarius Samples

EAHCP = Edwards Aquifer Habitat Conservation Plan
TWR = Texas wild rice

Ludwigia repens Competition Study

Submersed aquatic vegetation (SAV) in the San Marcos and Comal systems varies in terms of morphology, life history, and reproduction, and different species have been associated with varying densities of fountain darters. For this reason, a primary goal for the HCP has been seeking to establish or maintain vegetation that optimizes protected species' populations. For this study, a plant targeted for restoration—*Ludwigia repens*—was evaluated in terms of competition with non-native species. *Ludwigia* is associated with higher densities of fountain darter than two important non-native plants; *Hygrophila polysperma* (found in both river systems) and *Hydrilla verticillata* (identified only in the San Marcos). The objective of this project was to determine whether competition between *Ludwigia* and each of the non-native species was of importance in the early establishment period when new sprigs are planted in a cleared location. A second phase of the study continued to evaluate whether the plants were more or less vulnerable to competition after a period of established growth in the absence of the competing species.

While an earlier study had carried out initial, similar investigations in buckets (Doyle et al. 2003), the methods in this study better reflected the river systems by carrying out experiments under ambient conditions. Multiple locations were selected for the study to provide experimental replication. These locations represented different environmental conditions, such as high or low light and variable flow. Plants were grown in pots placed in the river and included control conditions for each species in addition to pots containing competing species with *Ludwigia*. The final report for this study provides greater details on the experimental design (Center for Reservoir and Aquatic Systems Research and BIO-WEST, Inc., 2015). Environmental conditions including depth, velocity, temperature, dissolved oxygen, pH, and photosynthetically active radiation were monitored along with response variables of maximum stem length, stem counts, and measurements of above- and below-ground biomass.

Across the experimental locations in pots without competition, *Ludwigia* exhibited strong growth, although *Hygrophila* and *Hydrilla* had the capacity to grow longer stems under some conditions. For the *Hygrophila* competition experiments in the early establishment phase of colonization, *Ludwigia* appeared capable of robust growth with only one site where competition was significant. It is possible that light availability is a key factor as the highest growth for *Ludwigia* appeared under high light conditions. The study's authors mention that the experimental locations generally had higher light conditions, so this leaves open a question as to whether the *Hygrophila* will have a competitive advantage under low light conditions. Contrasting with results from the earlier Doyle et al. (2003) study, the competitive experiments evaluating the impact of invasion by *Hygrophila* on established plants indicated that *Ludwigia* is not negatively impacted. Instead, *Hygrophila* appears to be negatively impacted by *Ludwigia*. **These results support renewed interest in *Ludwigia* as a species for restoration, especially in areas where *Hygrophila* is of concern. Insuring that this competitive advantage is communicated to the SAV modelers and incorporated into their efforts is also recommended.** In fact, this study should be highlighted as one where experimental results may be directly useful to the ecological modelers.

The *Hydrilla* experiments seem to have suffered from poor overall growth of this species, with declining biomass and evidence of mortality through the experiment, making the results difficult to interpret. The final report provides results indicating negative effects of *Hydrilla* on *Ludwigia* in this context, but it is not clear if the mortality events could impact *Ludwigia* grown in the same pots. **Additional consideration of the interactions between *Hydrilla* and *Ludwigia* is needed before conclusions are made or further application of this research occurs.**

A primary conclusion of this study is that *Ludwigia* can be planted successfully into unvegetated areas and is a good candidate for restoration. The results support this conclusion and the low competition with *Hygrophila* is particularly encouraging. Additional interpretation of the effect of experimental replicate location on growth suggest that environmental conditions are likely important factors to consider for restoration efforts, and may merit additional garden-style experiments investigating the relationship between light and flow on *Ludwigia* growth and colonization. The suggestions for further study listed at the conclusion of the final report are reasonable and would add to the knowledge available for improving restoration successes for this species. **Given the potential for *Ludwigia* to outcompete at least one of the non-native species, serious consideration should be given to using this species in the San Marcos system.** Although there has been mixed success with restoration of *Ludwigia* in the San Marcos system in the past, the results of this study suggest that this SAV species may be particularly valuable because it sustains both high fountain darter densities and a competitive advantage against non-native species that have been targeted for removal.

Comal Springs Riffle Beetle Population Occupancy Modeling

The objective of this study (Zara Environmental, 2015) was to develop a system-wide estimate of the population size of *Heterelmis comalensis* that would serve as a baseline population estimate for this endangered species. The study took a random survey site selection approach of 95 spring outlets for monitoring *H. comalensis* populations in October 2014, a period of extremely low flow. This approach identified all spring outlets in the entire system (300 sites) and then randomly selected a set for monitoring, avoiding issues of biased estimates related to site selection based on pre-existing knowledge about the potential state of occupancy. Monitoring relied on the established cotton-cloth lure approach from previous surveys. Each sampling event was separated by 72 hours to maintain independence among sampling periods. Using occupancy modeling approaches, significant combinations of covariates were used as predictors of population occupancy and then a N-mixture repeated count model was used to make system-wide estimates of beetle population abundance based on available habitat.

The beetle was detected in 22 of 95 spring outlets over three sampling time periods, with 101 adults and 36 larvae counted. Using the two modeling approaches, an estimate of a total of 741 beetles (90% CI 471-1284) was made for the entire system at the time of the survey. This is a very low population estimate for the entire Comal Springs system.

This project was a substantial effort, and it greatly expanded earlier modeling efforts that simply used wetted area to estimate riffle beetle population abundance in the Comal system. However, because the sampling period was during an extremely low flow period, an unknown number of additional spring outlets that had been previously identified by Norris and Gibson (2013) were either not flowing or had very reduced flow that prevented monitoring. When these springs are not flowing it is impossible to estimate beetle abundance, and there are no definitive data to suggest that when the springs dry up the beetles survive in the springs. Earlier accounts suggest that the beetles move deeper into the springs via wetted interstitial spaces; however, it is unknown whether beetles found in springs that begin to flow after a dry period are from the spring itself or have emigrated from other sources in the system. The dry springs of previously known populations could be one reason for the very low population estimate determined in the report. Additionally, this was not a mark and recapture study, and so it is unknown if the beetles counted during subsequent sampling events were repeat counts, new counts, or a mixture of recounts and new counts. Thus, while this study was an effort at a snap-shot estimate of the *H. comalensis* population in Comal Springs, the survey provides no information on the life history characteristics (e.g., synchrony, number of generations, growth rates) and how the population abundance changes over time and in response to flow: this is a critical aspect of the HCP. Importantly, because the cotton-lure method of sampling has not been calibrated to estimate densities, the estimates presented here may not accurately reflect true population densities at the surveyed spring outlets. This could be another explanation for the overall detection rate of only 51 percent in occupied sites. **New and innovative Applied Research projects should determine a reliable and defendable collection method for *H. comalensis*. Furthermore, a validation study that encompasses repeated sampling from the same and new spring outlets to account for potential life history and flow effects on the population estimates is highly recommended.**

This occupancy model was the most directed research effort to date for estimating the population abundance of the CSRB in the entire Comal Springs system. However, as indicated by the authors, there were several serious flaws to the study design, including the sampling approach. The occupancy report at best provides very limited information on the variation in CSRB among springs and seeps and, at worst, suggests that the CSRB populations are incredibly low and perhaps not being managed appropriately by the HCP. Fortunately, the issue of an appropriate sampling approach has become a focus of a 2016 Applied Research project (as discussed above).

Comal Springs Riffle Beetle Habitat Connectivity Study

The goals of this Applied Research project were threefold: (1) to examine water quality conditions and survival of the CSRB at two experimental facilities (USFWS San Marcos Aquatic Resource Center [SMARC] and the Texas State University Freeman Aquatic Building [FAB]); (2) to evaluate the potential of riffle beetle surrogate species; and (3) to test and describe components of upwelling and lateral habitat connectivity for the CSRB. As part of the last goal, the importance of organic material to CSRB behavior and movement was evaluated in laboratory experiments, as well as a field study using stable isotopes to determine CSRB food resources and to test two alternative sampling approaches to the cotton lure.

This project identified several important aspects of the CSRB biology, namely that surrogate species are unlikely to be useful for understanding the CSRB environmental tolerances and habitat occupancy, that passive sampling using pit traps and pumping are not viable alternative sampling strategies, and that both leaf and wood organic matter is important to the habitat and biology of the CSRB. The latter finding suggests an alternative sampling approach as well as identifying important habitat characteristics that could be used in future modeling efforts. That is, if terrestrial-derived organic materials are important to the CSRB as a food source and as habitat, then the riparian plant community could play a role in affecting CSRB habitat and population structure. Additional studies will be required to vigorously test these potential impacts to modeling CSRB populations; it is evident from the 2016 Applied Research projects that this study has provided a foundation for future studies. The Committee commends this effort to fund new projects to build upon these informative results.

An interesting finding from this study was that the Pecks's Cave amphipod (*Stygobromus pecki*) could be a predator of other invertebrates in these interstitial habitats. If this finding can be confirmed and *S. pecki* is determined to be a predator (or facultative predator) of the CSRB, new studies on the importance of predation on CSRB populations would be warranted.

The first goal of the project highlighted issues of using laboratory experiments to make broad inferences about field conditions. While the few water quality variables that were measured (temperature, conductivity, dissolved oxygen) at each facility were the same, there was nearly 100 percent mortality of the CSRB at the FAB facility, for reasons that remain unknown. This finding reveals that there are significant dimensions of CSRB biology that are not yet understood that may be critical to the survivorship of this endangered species. This finding also suggests that past, current, and future experiments using these laboratory chambers may not represent how the species responds to environmental change in nature, creating a degree of uncertainty as to how these data can be used in monitoring and modeling the CSRB. **The Committee recommends that additional studies be conducted to identify the source of mortality at the FAB facility, since doing so would likely reveal important factors that are necessary for structuring and maintaining CSRB populations. A second recommendation would be to validate key laboratory experiments like the one in this connectivity study using creative field studies where variables can be manipulated.**

Algae and Dissolved Oxygen Dynamics of Landa Lake and the Upper Spring Run

The goal of this Applied Research project was to characterize the composition, spatial and temporal distribution, and ecological consequences of benthic algae turf mats and floating vegetation mats in Landa Lake and Upper Spring Run (BIO-WEST, Center for Reservoir and Aquatic Systems Research, and Aqua Strategies, 2015). The impetus for this study was the observation that the turf and floating vegetation mats were abundant in Landa Lake and Upper Spring Run during 2013 and 2014 (which were years of low flow) and there was concern that these mats could affect the distribution and abundance of fountain darter habitat through the mats' effects on SAV and dissolved oxygen concentration.

The development of benthic algal mats was monitored during the summer of 2015 in a series of permanent transects established throughout Landa Lake and Upper Spring. In addition, an experiment to assess the effects of benthic algal mats on other aquatic vegetation was attempted by planting *Ludwigia* and bryophytes in areas with and without benthic algal mats. To assess the effects of floating vegetation mats on dissolved oxygen concentrations, miniDOT oxygen sensors were placed in areas with and without floating mats and at different depths in the water column. Finally, a model was developed to assess the relative importance of reaeration, water column algae and macrophytes on dissolved oxygen concentration and dynamics.

Due to heavy spring rain events, the summer of 2015 had slightly larger than average flows. Perhaps because of the higher flow rates, both the benthic turf mats and the floating vegetation were not as prominent as in the prior two years of low flow. Therefore, although the study was originally planned to assess the development and effects of these mats during low flow years, the study actually characterized mat development during normal flow conditions. Nevertheless, the study provided useful baseline information on the relative importance and ecological effects of benthic and floating mats.

Benthic algal mats, which were largely made up of *Spirogyra* and *Cladophora*, rarely co-occurred with bryophyte mats. The transplant experiment to test whether the benthic mats caused reductions in bryophytes and other macrophytes was not definitive due to the difficulty of establishing an area of persistent benthic algal mat occurrence, but suggested there might be a negative effect between benthic algal mat abundance on other aquatic vegetation. The floating vegetation mats appeared to have only minor effects on dissolved oxygen, but the modeling results suggested that if the floating mats cover more than 25 percent of the lake surface area the mats could reduce dissolved oxygen in the water column by reducing reaeration rates. Not surprisingly, diurnal dissolved oxygen concentrations at the benthic surface had greater amplitude than those higher in the water column. The only evidence of minimum dissolved oxygen levels below 4 mg/L were in stagnant areas and within bryophyte mats, which fountain darters could likely avoid.

In general this study was well conceived, but the higher than average flows limited its usefulness to understand how benthic algal and floating vegetation/debris mats influence fountain darter habitat through effects on macrophytes and dissolved oxygen during low flow conditions. MiniDOTs oxygen sensors give great flexibility in monitoring spatial patterns in dissolved oxygen concentrations and should be integrated into the water quality and biological monitoring plans.

Dissolved Oxygen Management in Landa Lake

Ensuring adequate levels of dissolved oxygen in habitats important to fountain darter and other species of concern is important to meeting the goals of the HCP. Because sonde measurements of dissolved oxygen in Landa Lake have occasionally showed dissolved oxygen below the somewhat arbitrary regulatory threshold of 4 mg/L, a mitigation measure of aerating Landa Lake to increase dissolved oxygen concentrations and/or removing floating vegetation/debris mats is being considered.

Preliminary tests of an aeration system occurred in the summer and fall of 2015. The report *Supplemental Dissolved Oxygen Evaluation in Landa Lake* (BIO-WEST, 2015) was produced to report on aeration tests conducted in September 2015. As described above, an additional study on the effects of floating vegetation/debris mats on dissolved oxygen concentrations in Landa Lake was completed in 2015 as part of the Applied Research Program.

In September 2015 an overnight aeration test was done in Bleiders Creek, up flow from Landa Lake. This area was selected because relatively stagnant, low-flow conditions were present, mimicking the conditions that might occur in areas of Landa Lake during low spring flow conditions. Two aeration diffusers were operated over the course of one night, and numerous dissolved oxygen measurements were made using handheld sensors and sondes at multiple points surrounding the diffusers. Results showed relatively little effect of the diffusers on dissolved oxygen concentrations, raising dissolved oxygen concentrations by approximately 0.5 mg/L. These results were then used in model calculations of the potential effectiveness of aeration on the larger Landa Lake. Results of these calculations showed that it would take about 160 diffusers spaced 30 feet apart to increase the ambient dissolved oxygen concentration by 1 mg/L. The limited effectiveness of the diffusers is largely due to the shallow depth of the lake.

The EAA deployed multiple miniDOT oxygen sensors in Landa Lake and Upper Spring Run during the summer of 2015 to monitor spatial and temporal changes in dissolved oxygen concentrations. This is a relatively low cost and effective way to monitor oxygen levels, and this monitoring should be continued as part of a routine integrated water quality and biological monitoring program. The EAA can use the knowledge gained during 2015 to identify the key locations in Landa Lake that will serve as indicators for the entire lake system.

Results of the 2015 and previous studies suggest that low concentrations of dissolved oxygen are not a widespread problem in Landa Lake and Upper Spring Run except for a few isolated locations during stagnant periods of low flow. It is likely that fountain darters and other species of concern can move to avoid these areas of low oxygen concentration. Furthermore, the 2015 study on effects of aeration on dissolved oxygen concentrations (BIO-WEST, 2015) showed that aeration had only minimal effects, raising dissolved oxygen concentrations less than 1 mg/L. **Therefore, the Committee recommends that aeration not be used routinely as a mitigation measure.** In an emergency situation, if dangerously low levels of dissolved oxygen persist even in the deepest areas of Landa Lake, then using aerators in a small area of Landa Lake to create a small refuge of higher dissolved oxygen water should be considered. If floating mats cover more than 25 percent of the surface of Landa Lake and dissolved oxygen concentrations decrease, then manual breaking up and removal of the floating mats should be considered as a mitigation measure. **The Committee further recommends that monitoring the dissolved oxygen concentrations using the miniDOTs in selected areas of Landa Lake and Upper Spring Run be incorporated into an integrated water quality and biological monitoring program.**

FUTURE OF THE APPLIED RESEARCH PROGRAM

The Applied Research Work Group was formed in 2015 subsequent to the release of NRC (2015). The charge to this work group was two-fold: (1) determine if additional Applied Research studies are needed, and (2) develop a research plan that prioritizes the numerous studies that have been recommended by the National Academies' Committee, the Science Committee, the Implementing Committee, and independent subject matter experts.

One of the activities of the Applied Research Work Group was to identify categories of research. This was accomplished and led to the following five categories:

1. **Conservation Measures:** Assessing the holistic practical benefits of HCP conservation measures to the species, and the effectiveness

of the conservation measures in achieving biological objectives and goals.
2. **Standard Sampling Methods:** Establishing reliable sampling methods for the species to ensure they permit evaluation of trends over time, including standardization as an important goal; and that they are consistent with biological objectives and goals.
3. **Habitat Quality, Quantity, and Requirements:** Evaluating the habitat requirements of the species, including the assessment of whether habitat is of sufficient quality and quantity, and validating HCP assumptions related to habitat, consistent with biological objectives and goals.
4. **System Memory/Disturbance Ecology:** Measuring the effects of disturbance (e.g., drought, scouring floods, etc.) on the system, and the response (i.e., resilience and/or resistance) of the system post-disturbance as it relates to biological objectives and goals.
5. **Data:** Data management considerations relevant to existing and future data to be collected, as well as applications for analysis of existing data relevant to biological objectives and goals.

The future projects that the Applied Research Work Group decided would fill out the remainder of the program's time are listed in Table 5-5.

In general, the Committee is supportive of the Work Group's efforts to identify priority areas for the Applied Research Program and to plan projects through 2019. The 2017 project to establish better relationships between the fountain darter and the different species and coverages of SAV (including *Ludwigia*) in both systems is critically important. This will be essential information to have when removing non-native species to insure that take is minimized. Research to better understand the life history of listed species and identifying effective sampling techniques rightfully deserves high priority. This knowledge underpins efforts to assess the ecological status and trends of the Comal Springs riffle beetle, Comal Springs dryopid beetle, and Peck's Cave amphipod. This research is needed but will be difficult and may require multiple years to be successful. The EAA should be prepared to invest in additional research projects in this area that span multiple years, if necessary.

Starting in 2018, the EAA appears to be using the Applied Research Program as a mechanism to assess the effectiveness of minimization and mitigation measures such as removal of exotic species, SAV restoration, and sediment control. While monitoring the effectiveness of these measures is critical, it is not clear that this monitoring should be part of the Applied Research Program. Ecological effects resulting from these minimization and mitigation measures are likely to play out over the long term rather than in a single year. Therefore, monitoring to assess the effectiveness of

TABLE 5-5 Projects for the Applied Research Program 2017-2019

Project Title	Year	Committee Comments
Continuation of CSRB life history study	2017	Continuation of the study begun in 2016.
SAV as FD habitat (shelter, prey habitat)	2017	This is a key metric worthy of continued efforts. Quantifying the relationship between a given SAV species and coverage with FD will serve to inform M&M measures and will help in estimates of FD density based on vegetative cover.
Effects of sedimentation on SAV, FD and CSRB*	2017	The RFP focuses specifically on the CSRB.
Comal Springs dryopid beetle quantitative sampling techniques	2017	Previously unsampled organism.
Statistical analysis of data × 2*	2017	Two projects will develop additional study questions to further explore biological objectives and statistically analyze existing EAHCP data concerning system memory/disturbance ecology and species-specific questions.
Peck's Cave amphipod quantitative sampling techniques	2018	Previously unsampled organism.
Evaluate success of SAV restoration and TWR enhancement (coincides with 5-yr SAV mapping)	2018	Performance monitoring of an M&M. Future efforts need to also be mindful of evaluating success as normalized to effort and potential available habitat.
Confirm species-specific Tables 4-1, 4-21 in the HCP	2018	Performance monitoring of an M&M. Tables 4-1 and 4-21 list the numbers of FD found in an area of SAV, for Comal and San Marcos, respectively. Could be similar to 2017 study.
Evaluate success of flow-split management	2018	Performance monitoring of an M&M
Contingency slot	2018	
Evaluate success of removal of invasive animal species and reduction of introduction	2019	Performance monitoring of an M&M
Evaluate success of Sessom Creek sand bar removal and sediment removal efforts	2019	Performance monitoring of an M&M
Contingency slot	2019	

*The Committee's reading of these RFPs finds it will be very hard for those unfamiliar with the system to respond to this RFP, for a number of reasons.
SOURCE: EAHCP ARWG (2015).

the minimization and mitigation measures needs to be ongoing through the lifetime of the HCP. As discussed in Chapter 4, the Committee recommends that such monitoring should be integrated into the biological and water quality monitoring programs rather than done in one-year studies through the Applied Research Program.

CONCLUSIONS AND RECOMMENDATIONS

The Committee applauds the changes made by the EAA regarding the procedures to identify, solicit, and review the projects in the Applied Research Program. The program as modified should be continued and could be expanded to facilitate additional multi-year studies in the future. It has the potential to provide data and understanding of basic processes that will help inform implementation of the mitigation and minimization measures as well as the development of the ecological models. To encourage more involvement of more outside experts, the EAA should look for ways to ease barriers to participation in the Applied Research Program.

The Committee is supportive of EAA's attempts to develop an effective database management system that will provide data storage, curation, and access into the future. Resources for ongoing data management activities will need to be allocated throughout the lifetime of the HCP.

Monitoring the effectiveness of minimization and mitigation measures such as removal of exotic species, sediment control, and riparian conservation should be done through integration into the existing biological and water quality monitoring programs, rather than through one-off studies conducted through the Applied Research Program.

Modeling efforts should become more integral to consideration of future Applied Research projects. Projects in the Applied Research program can provide data and information to help design model scenarios, to improve parameter estimation and model formulation, and to enable model calibration and validation. For example, the NRC (2015) and NASEM (2016) recommendation that nutrients be considered in the ecological sub-model of SAV would be easier to implement with nutrient data collection and more explicit consideration of nutrients in Applied Research projects related to SAV.

REFERENCES

BIO-WEST. 2014a. Fountain Darter Movement under Low flow Conditions in the Comal Springs/River Ecosystem. Final Report. October 30, 2014.
BIO-WEST. 2014b. Effect of Low-Flow on Riffle Beetle Survival in Laboratory Conditions. Final Report. November 14, 2014.
BIO-WEST. 2015. Final report. Supplemental Dissolved Oxygen Evaluation in Landa Lake.

BIO-WEST. 2016. Evaluation of the Life History of the Comal Springs Riffle Beetle. EAHCP Proposal NO. 147-15-HCP.
BIO-WEST and Baylor University. 2013. Edwards Aquifer Habitat Conservation Plan 2013 Applied Research. Final Version. November 2013.
BIO-WEST and Texas State. 2015. Comal Springs Riffle Beetle Habitat Connectivity Study. December 14, 2015.
BIO-WEST, Center for Reservoir and Aquatic Systems Research, and Aqua Strategies. 2015. Algae and Dissolved Oxygen Dynamics of Landa Lake and the Upper Spring Run.
Center for Reservoir and Aquatic Systems Research and BIO-WEST, Inc. 2015. Final Report for *Ludwigia repens* Competition Study. Edwards Aquifer Authority Contract #14-727L
Doyle, R. D., M. D. Francis, and R. M. Smart. 2003. Interference competition between *Ludwigia repens* and *Hygrophila polysperma*: two morphologically similar aquatic plant species. Aquatic Botany 77:223-234.
Doyle, R., S. Hester, and C. Williams. 2014. Edwards Aquifer Authority 2014 Ecomodeling: Vegetation Percent Cover to Biomass. Report of Research Activities. November 25, 2014.
EAA. 2015. Applied Research Selection Process.
EAHCP ARWG. 2015. Report of the 2015 Applied Research Work Group. October 16, 2015. Edwards Aquifer Habitat Conservation Plan Applied Research Work Group.
Gibson, R., C. Norris, and P. Diaz. 2013. Determination of Limitations of Comal Springs Riffle Beetle Plastron Use during Low-Flow Study. Proposal #124-13-HCP. August 28, 2013.
NASEM (The National Academies of Sciences, Engineering, and Medicine). 2016. Evaluation of the Predictive Ecological Model for the Edwards Aquifer Habitat Conservation Plan: An Interim Report as Part of Phase 2. Washington, DC: The National Academies Press.
NRC (National Research Council). 2015. Review of the Edwards Aquifer Habitat Conservation Plan: Report 1. Washington, DC: The National Academies Press.
Norris, C., and R. Gibson. 2013. Distribution, Abundance and Characterization of Freshwater Springs Forming the Comal Springs System, New Braunfels, Texas. Report prepared for Texas Parks and Wildlife Department.
Nowlin, W., and D. Hahn. 2015. Evaluation of the Trophic Status and Functional Feeding Group Status of the Comal Springs Riffle Beetle. EAHCP Proposal NO. 148-15-HCP.
Nowlin, W., and B. Schwartz. 2015. Evaluation of Long-Term Elevated Temperature and Low Dissolved Oxygen Tolerances of the Comal Springs Riffle Beetle. EAHCP Proposal NO. 146-15-HCP.
Texas State University. 2016. Suspended Sediment Impacts on Texas Wild Rice and other Aquatic Plant Growth Characteristics and Aquatic Macroinvertebrates. September 2016.
Texas State University and BIO-WEST. 2014a. Effects of Low Flow on Fountain Darter Reproductive Effort. Final Report. October 2014.
Texas State University and BIO-WEST. 2014b. Effects of Predation on Fountain Darters Study. Final Report. October 2014.
Zara Environmental. 2015. Comal Springs Riffle Beetle Occupancy Modeling and Population Estimate within the Comal Springs System, New Braunfels, Texas. Prepared by Zara Environmental LLC and submitted on 23 March 2015.

6

Mitigation and Minimization Measures

INTRODUCTION

The federal Endangered Species Act's incidental take permit (ITP) provisions require minimization and mitigation (M&M) measures. First, during the application process, the applicant for an ITP must specify "what steps the applicant will take to minimize and mitigate [the impacts of its incidental taking of ESA-listed species] and the funding that will be available to implement such steps." Second, before issuing the ITP, the U.S. Fish and Wildlife Service (FWS) must find that "the applicant will, to the maximum extent practicable, minimize and mitigate the impacts of such a taking. . . ." Thus, M&M measures are an integral part of the ITP process and requirements.

The Habitat Conservation Plan (HCP) lists 38 M&M measures that have an associated budget. Table 6-1 summarizes these measures, giving the cost of measure implementation during the first half of the ITP (Years 1 to 7). The table also relates the Committee's opinion about the species likely to directly benefit from the measure and the purpose of the measure. It should be noted that there are additional measures under consideration that have been the subjects of Applied Research projects (e.g., dissolved oxygen management). Thus, the table does not include all potential measures being implemented or under consideration, only those found in the HCP with an associated budget.

The purpose of this chapter is to review key M&M measures as well as to identify approaches that might maximize their effectiveness. The specific M&M measures reviewed in this chapter include the following:

TABLE 6-1 M&M Measures Found in the HCP

M&M Measure (HCP Section)	Total Cost Years 1-7	Target Species	Purpose
Flow Protection Measures			
Critical Period Management Stage 5 (5.1.4)	NA	All	Reduce all pumping during critical low flows
Aquifer Storage and Recovery (ASR) (5.5.1)	$33.3 mil (leases) $15.3 mil (O&M)	All	Reduce SAWS pumping
Regional Water Conservation Program (5.1.3)	$11.3 million	All	Reduce demand of SAWS and other participants
Voluntary Irrigation Suspension Program Option (5.1.2)	$29.2 million	All	Reduce irrigator pumping
San Marcos System			
Texas wild rice enhancement and restoration (5.3.1, 5.4.1)	$1.05 million	TWR	Maintain TWR in the system
Sediment removal at Sewell Park (5.3.6 and 5.4.4)	$650,000	TWR, SAV	Prevent sediment from smothering TWR
Aquatic vegetation restoration (non-native removal and native reestablishment) and maintenance (5.3.8, 5.4.3, 5.4.12)	$975,000	SAV, TWR	Provide FD habitat; remove competition from TWR
Management of floating vegetation mats and litter removal (5.3.3 and 5.4.3)	$560,000	FD, TWR, SAV	Reduce light attenuation and promote SAV/TWR growth. Prevent DO depletion
Non-native animal species control (5.3.5. 5.3.9, 5.4.11, 5.4.13)	$245,000	FD, SAV	Prevent predation of FD and destruction of SAV
Sessom Creek sand bar removal (5.4.6)	$100,000	TWR	Prevent sediment from smothering TWR; limit hydraulic changes in river
Low impact development/BMPs (5.7.3)	$2 million	All	Prevent contaminants from entering river systems via surface runoff
Recreation control in key areas (5.3.2, 5.4.2)	$336,000	TWR, SAV	Prevent physical damage to species

TABLE 6-1 Continued

M&M Measure (HCP Section)	Total Cost Years 1-7	Target Species	Purpose
Restoration of riparian zone with native vegetation (5.7.1)	$220,000	TWR, SAV, CSRB	Prevent bank erosion
Bank stabilization/permanent access points (5.3.7)	$120,000	TWR, SAV, CSRB	Prevent bank erosion
Household hazardous waste program (5.7.5)	$210,000	All	Prevent contaminants from potentially entering the rivers by removing them from the watershed
Comal System			
Old Channel Environmental Restoration and Protection Area (ERPA) (5.2.2.1)	$1.2 million	SAV	Promote health of prime FD habitat
Flow-split management (5.2.1)	$210,000	SAV, FD	Control hydraulics into Old Channel
Landa Lake and Comal River aquatic vegetation restoration and maintenance (5.2.2)	$845,000	SAV	Provide FD habitat
Non-native animal species control (5.2.5, 5.2.9)	$645,000	FD, SAV	Prevent predation of FD and destruction of SAV
Decaying vegetation removal program (5.2.4)	$840,000	FD, SAV	Reduce light attenuation and promote SAV growth. Prevent DO depletion
Riparian improvements and sediment removal specific to the CSRB (5.2.8)	$325,000	SAV, CSRB	Prevent bank erosion, sediment accumulation, and potential siltation of CSRB habitat
Gill parasite control and non-native snail removal program (5.2.6)	$725,000	FD	Prevent FD disease
Restoration of riparian zone with native vegetation (5.7.1)	$800,000	SAV, CSRB	Prevent bank erosion

continued

TABLE 6-1 Continued

M&M Measure (HCP Section)	Total Cost Years 1-7	Target Species	Purpose
Prohibition of hazardous materials route (5.2.7)	$10,000	All	Prevent contaminants from entering the river systems by routing them away from rivers
BMPs for stormwater control (5.7.6)	$300,000	All	Prevent contaminants from entering river systems via surface runoff
Incentive program for Low Impact Development (LID) (5.7.6)	$700,000	All	Promote BMPs, which can prevent contaminants from entering rivers via surface runoff
Household hazardous waste program	$210,000	All	Prevent contaminants from potentially entering the rivers by removing them from the watershed
Common to both systems, not habitat restoration			
Biomonitoring both systems (6.3.1)	$2.8 million	All those monitored	Assess listed species populations and some WQ parameters
Water quality monitoring both systems (5.7.4)	$1.4 million	All	See trends in WQ parameters; compare to WQ in biomonitoring
Development of a mechanistic ecological model (6.6.3)	$950,000	FD, SAV	Predict the responses of listed species to changes (e.g., in flow)
Applied environmental research at the USFWS National Fish Hatchery and Training Center Refugia (6.3.4)	$4.75 million	All those under study	Gain new knowledge of listed species; update ecomodel
Science Review Panel (NAS)	$550,000	All	Provide ongoing advice
Improve Groundwater Model	Budgeted prior to and outside the HCP	All	Better predict spring flow and well responses to changes in climate, management options, etc.
National Fish Hatchery and Training Center Refugia	$11.75 million	All those included	A repository of genes/organisms to seed springs if populations decline

Note that there were other M&M measures in the HCP (labeled as "other") for which no budget was provided. Also, there are activities, such as dissolved oxygen management, that are not explicitly listed in the HCP as M&M measures.

- Submersed aquatic vegetation (SAV) restoration/invasive plant removal in both the Comal and San Marcos systems
- Sediment removal at specific locations
- Dissolved oxygen management in Landa Lake
- Voluntary Irrigation Suspension Program Option (VISPO)
- Regional Water Conservation Program (RWCP)
- Stage V Critical Management Period
- Aquifer Storage and Recovery (ASR)

These M&M measures were selected because of their importance to reaching the biological goals and objectives of the HCP, or because they were specifically identified for review by the Edwards Aquifer Authority (EAA) as a result of uncertainties about their effectiveness or implementation.

Minimization and Mitigation Measures Background and History

Litigation in 1993 focused on reduced spring and stream flow as the primary source of impacts to Edwards Aquifer ESA-listed species, *Sierra Club v. Babbitt* (No. MO-91-CA-069, U.S. Dist. Ct., W.D. Texas), and the Edwards Aquifer Habitat Conservation Plan (HCP) and ITP maintain that focus (HCP 4-36). Specifically, the U.S. District Court for the Western District of Texas in 1993 ordered the FWS to establish five flow limits: "(1) the spring flow levels at which take of fountain darters and Texas blind salamanders begins at Comal and San Marcos Springs, (2) spring flows necessary to avoid appreciable diminution of the value of critical habitat of any listed species; (3) the spring flow at which Texas wild rice begins to be damaged or destroyed; (4) the minimum spring flow to avoid jeopardy for the fountain darter, San Marcos gambusia, San Marcos salamander, and Texas blind salamander; and (5) the spring flow levels at which take of San Marcos gambusia and the San Marcos salamander begins at San Marcos Springs" (HCP 4-36).

The FWS acknowledged in 1993 that it lacked adequate data for both its "take" and its "jeopardy" determinations. Nevertheless, the flow minimums it submitted to court based on its best professional judgment were as shown in Table 6-2 (adapted from HCP 4-37, Table 4-28), which the FWS characterized as "conservative."

However, the FWS's 1993 report to the U.S. District Court for the Western District of Texas also suggested other habitat protection efforts that became the inspiration for M&M measures. For example, the "FWS found that flow levels at Comal Springs could be reduced to 60 cfs for short time periods during certain times of the year without jeopardizing the continued existence of the fountain darter if a 'very effective' program to control the giant rams-horn snail was in place and if there was the abil-

TABLE 6-2 Flow Minimums for Edwards Species for Various Purposes

Species	Level of Flow for ESA "Take"	Level of Flow for ESA "Jeopardy"	Level of Flow for ESA Adverse Modification of Habitat
Fountain Darter in Comal	200 cfs	100 cfs	100 cfs
Fountain Darter in San Marcos	60 cfs	50 cfs (San Marcos Spring flow)	150 cfs
San Marcos Gambusia	100 cfs	100 cfs	60 cfs
San Marcos Salamander	50 cfs (San Marcos Spring flow)	N/A	100 cfs
Texas Blind Salamander	100 cfs	60 cfs	N/A
Texas Wild Rice ("damage or destruction" standard from ESA for plants)	100 cfs	100 cfs	100 cfs

ity to control the timing and duration of low spring flows" (HCP 4-37). The FWS "also found that short-term reductions in flow levels below 100 cfs might avoid jeopardy for Texas wild rice, if: (1) exotic species (e.g., nutria) could be effectively controlled, (2) an aquifer management plan is implemented to control timing and duration of lower flows, and (3) the distribution of the species is improved throughout its historic range. FWS, however, did not specify what flow levels might be acceptable if those conditions were satisfied" (HCP 4-37). Thus, also from the beginning of the ESA process, management of the Edwards Aquifer system focused on habitat improvements that could reduce the minimum flow requirements.

The FWS's Final Environmental Impact Statement (EIS, December 2012, http://www.fws.gov/southwest/es/Documents/R2ES/EARIP_HCP_FEIS.pdf) for the Edwards Aquifer HCP and ITP compared four proposals for managing the system. Specifically, the Environmental Impact Statement (EIS) evaluated four alternatives: (1) a "no action" that assessed the status quo if no ITP were issued and no streamflow protections were instituted (required by the National Environmental Policy Act as the baseline alternative); (2) the proposed EARIP HCP; (3) an expanded Aquifer Storage and Recovery (ASR) effort, with its associated infrastructure; and (4) highest critical management period pumping restrictions, which would require an

85 percent reduction in pumping during drought conditions, to 85,800 acre-feet per year (105.791 million m³ per year) (EIS ES-iii). Only Alternatives 2 and 3 required an HCP and ITP. Alternative 2 was the FWS's preferred alternative:

> Though the activities covered under this alternative could generate impacts to covered species, implementation of the proposed HCP is expected to contribute to recovery of the listed species and ensure their survival during conditions equivalent to those experienced during the [drought of record]. The anticipated cost of implementing Alternative 2 has been estimated to total $261.2 million over the 15-year life of the permit. Funding obligations associated with implementing the proposed HCP could have some negative economic impacts, though the certainty provided by an ITP ensuring continued use of the Edwards Aquifer is expected to be an overall benefit to the regional economy. The EARIP HCP is the alternative that minimizes negative effects to both the natural and human environment to the greatest extent, and is the Service's preferred alternative (EIS ES-iv).

In contrast, Alternative 1 would not protect either the species or the local economy from an extended drought; Alternative 3 would cost $439 million to $1.16 billion; and "the indirect and cumulative effects resulting from the proposed pumping restrictions and developing alternative water sources for human use under Alternative 4 would be expected to have significant negative economic impacts throughout the region" (EIS ES-iv).

The four alternatives also differed radically in their M&M measures. Implementation measures common to all four alternatives included the EAA's groundwater withdrawal program and permit administration (mandated under Texas law by the EAA Act); the City of New Braunfels' management of golf course diversions, spring-fed pool diversions, and boat operations on Comal River and Landa Lake, along with infrastructure maintenance and repair, litter collection, and floating vegetation management; and the City of San Marcos's management of boat operations on the San Marcos River and its infrastructure maintenance and repair (EIS 2-2 Table 2-1). However, as summarized in the EIS, only Alternative 2 offered multiple M&M measures to protect and manage spring flow at Comal and San Marcos Springs; Alternative 3 relied solely on large ASR projects, while Alternatives 1 and 4 did not offer any M&M measures (EIS 2-4 to 2-5, Table 2-2). Alternatives 2 and 3 proposed identical measures to minimize and mitigate impacts to the spring ecosystems (EIS 2-5 to 2-6, Table 2-2). Thus, collectively, Alternative 2 proposed the greatest number of minimization and mitigation measures to protect against the most varied types of impacts, and it was implemented.

REVIEW OF SELECT M&M MEASURES

Vegetation Restoration/Invasive Removal

Minimization and mitigation measures related to vegetation are carried out by the Cities of New Braunfels and San Marcos as well as Texas State University. These activities are largely related to habitat protection or enhancement for the fountain darter and focus on non-native SAV removal, restoration of native SAV species with special attention to the endangered Texas wild rice, and removal of decaying vegetation and maintenance of newly restored sites.

Of interest to the EAA has been understanding the relative differences in restoration and non-native removal techniques applied to the San Marcos and Comal Rivers. This is a valid comparison to make, as lessons learned from each system might be used to improve the effectiveness of vegetation M&M measures in both. In both systems, implementation of these M&M measures is logistically complicated, requiring trained crew who can be relied upon for hard manual labor, have SCUBA or snorkeling experience, and possess scientific technical skills and conscientious quality control habits. Timing must be coordinated between cultivation of plants for use in restoration, and collection of data to document success rates, all the while being mindful of observational data as new techniques or locations are pulled into the restoration effort. In this context, it is likely that each group managing implementation of the vegetation M&M measures has designed a strategy that is uniquely optimized to their organizational and management structure for field and scientific teams. There may be the perception that there is little room for experimentation or adaptation of another's methods as it would be disruptive in a way that would have unintended consequences on the overall productivity of the teams working in each system.

Fortunately, the two teams that have been implementing vegetation M&M measures in the Comal and San Marcos systems have recently documented differences in technique and provided data to thoroughly and quantitatively evaluate success rates using metrics such as acres planted versus acres sustained (BIO-WEST and Watershed Systems Group, 2016). This effort to synthesize the datasets should serve as an example of how similar efforts related to other M&M measures or monitoring data might be carried out in the future. **Every time an M&M measure is implemented, there is a need to document whether it is working. This should be done not only for the first year of implementation, but periodically with a comprehensive synthesis of the monitoring data every five years or so that goes beyond the simple trends analyses found in the HCP annual reports.**

Recent restoration efforts in the Comal system involve *in situ* nursery areas and plugs of plants weighed down in trays that have been deployed across Landa Lake and the Old Channel. Plants used in the San Marcos system are grown at a facility at Texas State University, and planting occurs immediately following non-native vegetation removal. Removal of non-native plants has focused on *Hygrophila* in the Comal systems and *Hygrophila* and *Hydrilla* in the San Marcos system. Native plants for planting have included *Ludwigia, Sagittaria, Cabomba, Heteranthera,* and more recently *Potamogeton, Vallisneria, Justicia,* and bryophytes. *Justicia* and *Vallisneria* are not target species in the HCP, but their efficacy in restoration is being explored.

Overall, the BIO-WEST and Watershed Systems Group (2016) study evaluated existing vegetation coverage and presented the area of non-native plants removed versus the sustained planting coverage of native species. In the Comal system, this has resulted in 5,000 m^2 of *Hygrophila* removed since 2013. At the same time, 36,000 native plants were restored resulting in a total of 1,800 m^2 of sustained coverage. The definition of "sustained" takes into account the variable coverage from seasonal and inter-annual variability in the populations. In the San Marcos system, 1,800 m^2 of *Hygrophila* has been removed along with 3,400 m^2 of non-native *Hydrilla*. For the time frame evaluated in this study, it is interesting to note that total removal of non-native plants was approximately the same in the two systems (~5,000 m^2). The dollars invested in these two programs are also roughly comparable, suggesting that the efficiency of the two teams carrying out the work is about the same.

In terms of native plants restored to the two systems, BIO-WEST and Watershed Systems Group (2016) report a total of 36,000 plants restored to the Comal system, resulting in 1,800 m^2 of sustained growth. In the San Marcos system, 22,000 plants were restored, resulting in 700 m^2 of sustained coverage. In addition, the San Marcos system separates out their highly successful Texas wild rice program to report 30,000 total plants as associated with 3,600 m^2 of areal coverage by February 2016. The ratios of individual plants to resulting coverage in square meters is similar between the two systems (20:1 in the Comal and 31:1 in the San Marcos). The Texas wild rice ratio of eight plants for every resulting square meter of coverage is particularly impressive. This exercise of comparing planting effort to resulting coverage is useful, and we recommend continuing to compute ratios from data such as those reported in BIO-WEST and Watershed Systems Group (2016), further refining the data to be as species specific as possible. The goal of such an exercise should be to seek ways to lower this ratio so that restoration efforts are efficient. Of course, the mechanisms for a given ratio will be a function of a given species colonization potential, growth habits, survival rates, and reproduction.

Although an impressive amount of work has been to both remove non-native vegetation and restore natives, it must be emphasized that a primary driver for these M&M measures is to provide habitat for the fountain darter. In terms of areal coverage, the overall habitat balance from these efforts has been negative. In the Comal system, there is a net negative coverage of 3,200 m^2 from 2013 to February of 2016.[1] In the San Marcos system, there is a net negative restoration of SAV on the order of 4,500 m^2, although Texas wild rice has provided 3,600 m^2 of emergent vegetation. [It should be noted that coverage was adversely affected by a major storm event in October 2015, during which 30 percent of restored *Cabomba* plantings were lost and upstream erosion led to substantial burial of *Ludwigia* (Blanton and Assoc., 2016). This points to a potential area of coordination, where upstream erosion and stormwater runoff control measures may be needed to protect planting and sediment control efforts downstream.] The repercussions of these vegetative changes on the fountain darter population are not entirely clear because the data reported in BIO-WEST and Watershed Systems Group (2016) are only from the project areas, which make up only a subset of the total SAV coverage, and because different SAV species have different carrying capacities for fountain darter. In 2013, approximately 37,824 m^2 of total vegetation was documented in Landa Lake, while in the Old Channel Project Area there was 4,572 m^2.[2] So the net loss of SAV in the Comal system due to the vegetation M&M measures constituted about 7.5 percent of the total area. Unless the non-native SAV is poor fountain darter habitat (which it is not), **there is not enough new habitat from native plantings to maintain populations of fountain darter to balance non-native plant removal.** It may be unpalatable to consider non-native vegetation as fountain darter habitat, but the data used to develop the HCP indicate that both *Hydrilla* and *Hygrophila* serve this role. Ultimately, consideration of the vegetation M&M measures must encompass preservation of the fountain darter and Texas wild rice, the two species specifically targeted for protection in the HCP, as the primary end goal for any removal or restoration efforts.

Performance monitoring of the vegetation M&M measures should continue to follow the example laid out in BIO-WEST and Watershed Systems Group (2016). Useful metrics presented therein included number of plants planted, resulting sustained area, coverage of vegetation from baseline maps in 2013, and lessons learned regarding new species or techniques. Another important value to track is the difference between the area of non-native plants removed and the sustained native coverage (reported as m^2). This

---

[1] This calculation is based on 5,000 m^2 of *Hygrophila* removed and 1,800 m^2 of native SAV sustainably replanted in the Comal system, and on 3,400 m^2 of *Hydrilla* and 1,800 m^2 of *Hygrophila* removed and 700 m^2 of native SAV sustainably replanted in the San Marcos system (BIO-WEST and Watershed Systems Group, 2016).

[2] These data come from Table 1 of BIO-WEST and Watershed Systems Group (2016) and reflect the total project area minus the area of bare substrate.

net restoration value can then be compared with the baseline vegetation coverage for a given project reach (as we have done above for the Comal system) to determine the percent change in habitat availability. In order for this to be useful, there must be some effort to convert the areas of non-native removal and sustained coverage to fountain darter populations, using data for associated fountain darter densities as reported in the HCP or in the calculation of take.

Sediment Accumulation and Removal

Minimization and mitigation measures in support of the HCP include sediment removal at specific locations in the Comal and San Marcos Rivers. The primary goal of this effort is to eliminate sediment that has accumulated that may negatively impact existing SAV or may hinder colonization or survival of new SAV. This may be of direct concern (e.g., to Texas wild rice), or indirect, impacting fountain darters as a result of loss of habitat. The sediment has accumulated as a result of modifications to the river channel, dams, and urbanization and resulting alteration of stormwater runoff as well as natural processes.

Hydraulic suction is being used to remove the accumulated sediment. Divers first disturb the vegetation to drive biota away to minimize their collection during sediment removal. The nozzle of the vacuum is kept in the sediment substrate and not allowed to swing through the water column during the operation. Sensitive areas are also marked to ensure avoidance of vegetated areas of Texas wild rice. An observer is used to monitor the effluent for presence of listed species and all other biota, as well as for the safety of the diver. Sediment samples are sent to the Texas Commission on Environmental Quality (TCEQ) for contaminant testing per TCEQ requirements.

In the Comal River, one specific area of targeted sediment removal is a small island that has formed just behind the Springfed Pool and immediately downstream of Landa Lake. This sediment island continues to grow, has established destructive non-native vegetation, and has displaced/destroyed fountain darter habitat. Sediment is also of concern along Spring Reach 3 at the western end of the Landa Lake due to steep hillsides and development on that hillside. Efforts in this area are focused on minimizing sediment runoff into Landa Lake to reduce potential negative impacts on the Comal Springs riffle beetle.

In the San Marcos River, the City of San Marcos is removing sediment from the river bottom at various locations from City Park to IH-35. These areas include reaches of the river in City Park, Veramendi Park, Bicentennial Park, Rio Vista Park, and Ramon Lucio Park. In 2015, the focus was on areas downstream of Sewell Park, particularly the confluence of the San

Marcos River with Purgatory Creek. Dredging in Spring Lake and Sewell Park has also been considered, but no dredging took place in these areas in 2015. The method chosen for sediment removal is manpower intensive and slow and of questionable effectiveness. Between November 2014 and November 2015, an area of approximately 284 m^2 and volume of 85 m^3 of fine sediment was removed from the San Marcos River (Blanton and Assoc., 2016). This is modest compared to the goal of adding more than 1,000 m^2 of Texas wild rice in 2015.

There appears to be limited knowledge as to whether ongoing deposition will restore sediment loads in these areas. It is possible that the locations where fine sediments have accumulated tend to be natural depositional sites where continued deposition is likely to occur without substantial control of the upstream watershed. Bank pins and turbidity loggers could be used to evaluate sediment deposition where background knowledge is not currently available. Bank pins are low cost and have been used extensively in the Chesapeake Bay watershed to measure erosion.

High flow events simultaneously erode the sediment bed and remove the restored fauna. As mentioned previously, during a high flow event in October 2015, some 30 percent of restored *Cabomba* plantings were lost and subsequent deposition of gravel and sediment due to erosion of upstream areas led to substantial burial of *Ludwigia*. More established plantings were less impacted, but this illustrates that both replanting efforts and sediment removal efforts are at risk without more effective control of upstream erosion. In the Comal River the same flood led to only 10 to 15 percent loss of restored vegetation but led to an approximate 80 percent loss of bryophytes and to the accumulation of sediment and other debris in habitat areas.

It is too early to determine whether any of the current specific sediment removal efforts are having positive impacts on the river and the biota. All sediment removal actions should be coupled to monitoring efforts to demonstrate their efficacy. In addition to approaches such as bank pins to assess erosion, water depth and sediment accumulation should be monitored in areas being considered for sediment removal as well as post-removal. Maps of sediment accumulation should be prepared and monitored over time in the same manner that vegetation area is monitored over time. Monitoring is critical to the evaluation of the effectiveness of sediment removal efforts.

In summary, sediment removal activities can provide positive benefits in terms of enhancing SAV habitat and minimizing negative impacts on both vegetation and associated biota. However, the effectiveness of the current actions is unclear, especially in the San Marcos River where the sediment dynamics (including ongoing deposition) are either unknown or known to be controlled by erosion in upstream areas. The EAA has limited control over management of upland erosion. It is of little use to implement time-consuming sediment removal actions if the sediment is redeposited in subsequent storm events. This has occurred at the confluence of Purgatory Creek with the San Marcos River where a single storm event led to excessive sediment accumulation in an area that had undergone sediment removal. Similarly, efforts at Bicentennial Park are likely to have limited success because any sediment removal may be reversed by high flow events. A major sediment removal at the confluence of Sessom Creek with the San Marcos River has undergone substantial assessment, but the effort may be negated by sediment input from areas upstream of the control of EAA. **In general, sediment removal activities should be limited to areas where ongoing upland sources or natural stream dynamics will NOT lead to deposition of new sediment within a matter of years.**

Given these constraints, it may seem more appropriate to direct efforts toward controlling potential sediment sources than to attempt to remove sediment already in the river. However, it should be noted that the benefits of stormwater control measures tend to be local, and it can be difficult to install enough upstream measures to see an effect in the main channel.

Dissolved Oxygen Management in Landa Lake

Managing the dissolved oxygen status of Landa Lake is a mitigation and minimization measure, but it is not specifically called for in the HCP. Rather, an Applied Research project on dissolved oxygen in Landa Lake was conducted in 2015, the results of which are described in detail in Chapter 5. In summary, it appears that low concentrations of dissolved oxygen are not a widespread problem in Landa Lake and Upper Spring run except for a few isolated locations during stagnant periods of low flow. It is likely that fountain darters and other species of concern can move to avoid these areas of low oxygen concentration. Furthermore, the 2015 study on effects of aeration on dissolved oxygen concentrations (BIO-WEST, 2015) showed that aeration had only minimal effects, raising dissolved oxygen concentrations less than 1 mg/L. Therefore, the Committee recommends that aeration not be used routinely as a mitigation measure but be held in reserve to be used only in case of severe low oxygen conditions throughout all of Landa Lake. If floating mats cover more than 25 percent of the surface of Landa Lake and dissolved oxygen concentrations decrease, then manual breaking up and removal of the floating mats should be considered as a mitigation measure. The Committee further recommends that monitoring the dissolved oxygen concentrations using the miniDOTs in selected areas of Landa Lake and Upper Spring Run be incorporated into an integrated water quality and biological monitoring program.

Voluntary Irrigation Suspension Program Option

The Voluntary Irrigation Suspension Program Option (VISPO) is a program for holders of irrigation water rights who are willing to suspend use of all or a portion of their authorized pumping rights. Participants are financially compensated. The EAA determines each October 1st if the aquifer has declined to a level at or below 635 feet above mean sea level in the J-17 index well (which lies about 2 miles north of downtown San Antonio in Bexar County). If this occurs, program participants are required to suspend pumping for the following calendar year. The goal of this voluntary program is to enroll 40,000 acre-feet (49.32 million m^3) of permitted irrigation rights that will remain unused in years of severe drought. The program has two options: five-year, and ten-year.

As of April 2016, there was a combined total enrollment of 40,921 acre-feet, with 25,471 acre-feet enrolled in the 5-year program option and 15,450 acre-feet in the 10-year program option. The total enrollment amount exceeds the program goal contained in the HCP. The 5-year commitments enrolled in VISPO extend until December 2018. As shown in Table 6-1, the program will cost $29.2 million over the first 7 years of the HCP. Since the program's inception, participants were required to reduce their pumping in 2015 because the well J-17 was below the trigger elevation on October 1, 2014.

One drawback of the VISPO is that the trigger is based on the October 1 groundwater elevation. In the September time frame, groundwater elevations are typically beginning to recover from summertime lows as water demands decrease. However, the October 1 date does not necessarily reflect the lowest groundwater elevation, somewhat limiting its utility as a spring flow protection measure. As discussed in Chapter 2, the Committee recommends that Phase 2 of the HCP implement a Decision Support System to replace the triggers for the spring flow protection measures. The Decision Support System would incorporate additional information, such as modeling projections of future groundwater elevations, to determine when the spring flow protection measures would be triggered. For example, the 12-month outlook of the water levels at an index well would be modeled probabilistically, and a pre-determined action would be taken if there is reasonable probability that the water level will be at or below a critical value within that 12-month period.

Regional Water Conservation Program

The Regional Water Conservation Program (RWCP) offers incentives to municipalities to encourage water conservation in exchange for half of all conserved water to remain unpumped in the aquifer for 15 years. The goal of the program is to conserve 20,000 acre-feet (24.66 million m^3) of Edwards Aquifer pumping. For the target of 20,000 acre-feet conserved, 10,000 acre-feet would remain in the aquifer to sustain aquifer levels in support of continued spring flow. Conserved water that remains in the aquifer is held in a "groundwater trust." The other 10,000 acre-feet of conserved groundwater is available for pumping by the participating entity.

The RWCP currently includes activities focused on municipal water retailers. The program also provides assistance through low-flow toilet programs, leak detection, and other water use efficiency efforts.

The 2014 HCP Annual Report states, "To show that this measure is reasonably certain to occur, the EAA's goal was to obtain 'initial commitments' in the amount of 10,000 acre-feet/year in 2013. As conserved water is committed to the groundwater trust, the initial commitment water is to be returned to the committing entity. At present, the San Antonio Water System (SAWS), Texas State University, and the City of San Marcos have made initial commitments in the amount of 8,400 acre-feet" (Blanton and Assoc., 2015).

The January 2015 report of the Regional Water Conservation Program Work Group (EAHCP, 2015) included recommendations to increase participation in the RWCP. In 2015 SAWS committed to conserve 19,612 acre-feet by the year 2020 through a new leak-repair program, with one-half of the commitment in the leak-repair program going to the groundwater trust so that the water remains in the aquifer. The City of Uvalde and Universal City are also participating in the program. It is anticipated that by 2020 the RWCP will exceed the program's goal.

Stage V Critical Management Period

The Stage V Critical Management Period is an additional step beyond Stages I-IV that requires a pumping reduction of 44 percent from any Edwards Aquifer groundwater permit. The Stage V triggers for the San Antonio Pool are a monthly average 625 ft groundwater elevation at well J-17 or if the Comal Springs flow rate reaches specific rates. The trigger for the Uvalde Pool is 840 ft groundwater elevation at well J-27 (which lies in the city of Uvalde). The HCP framework is designed so that Stage V is triggered only when other measures have not proven effective in maintaining spring flow during drought conditions.

Through April 2016, Stage V has not been triggered for the San Antonio Pool. Figure 6-1 shows the J-17 monitoring well groundwater elevation for the last six years. Since January 2010, the lowest J-17 groundwater elevation observed was 625.9 feet above mean sea level on September 3, 2014. In 2015, the groundwater elevation in well J-27 recovered from low levels that triggered Stage V restrictions in the Uvalde Pool. Currently there is no Critical Management Period in effect in the region.

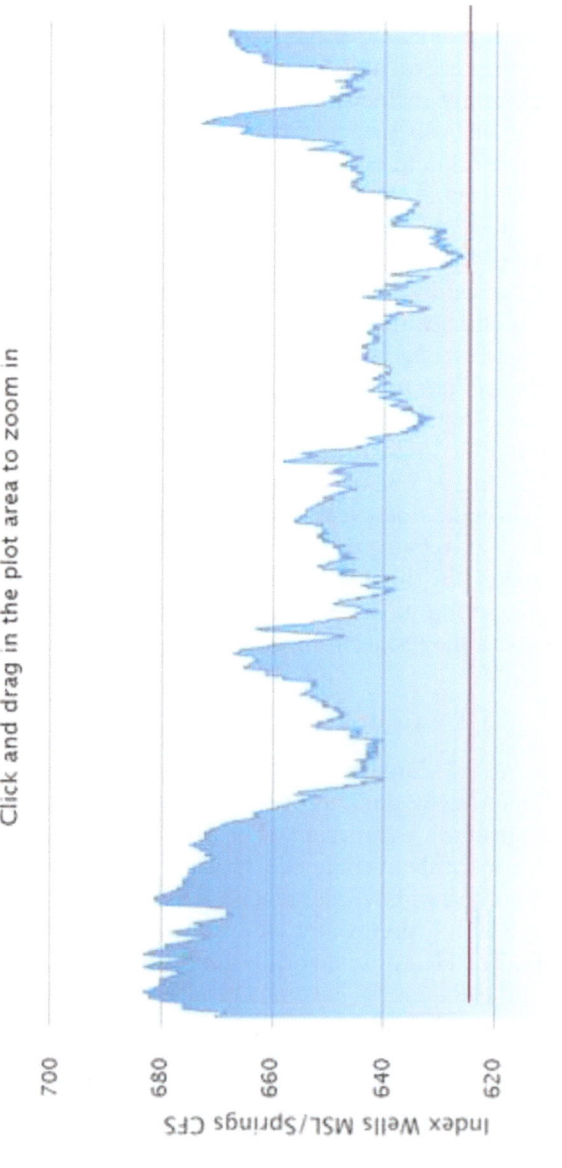

FIGURE 6-1 Water levels in J-17 well from 2010 to 2016. The bold line sits at 625 MSL.
SOURCE: www.edwardsaquifer.org (accessed 1-19-2016).

Aquifer Storage and Recovery

In August 2013, the EAA and SAWS developed an agreement for the use of the SAWS' Twin Oaks ASR in the HCP. The EAA-SAWS agreement terminates in March 2028. The EAA acquires leases of permitted Edwards Aquifer pumping, and this water is stored in the ASR system. When specified triggers are reached, SAWS will use water stored in the ASR as a base load supply in its area in the general vicinity of the springs, offsetting SAWS Edwards Aquifer demands.

Initially, the rate at which the EAA acquired leases to put water into the ASR lagged behind the target volume. However, in 2015 the EAA increased efforts to acquire leases. In 2015-early 2016, the EAA was managing a program that offers two methods for permit owners to participate: a leasing program with three lease options (10 years—$160 per acre-foot, 5 years—$140 per acre-foot, and 1 year—$120 per acre-foot) and a pooling program. The pooling program utilizes storage of unpumped water. On an annual basis, the pooling program estimates the volume of water that will be unpumped by pooling participants and provides notice to SAWS to store a portion of that water in the ASR. Participants in the pooling program are compensated $50 per acre-foot for their participation. The EAA has indicated that they will continue implementing the leasing program and pooling program through 2016 and make adjustments to the program going forward. For the ASR flow protection measure, the EAA is committed to acquiring 50,000 acre-feet (61.65 million m^3) of Edwards permits. As of mid-April 2016, the total leases were 27,015 acre-feet. In addition, in 2015 the EAA was able to provide SAWS with 500 acre-feet of pooled water through the pooling program.

Evaluation of Long-Term Reliability of Twin Oaks ASR

Two studies regarding the SAWS Twin Oaks ASR have been prepared under the supervision of Alan Dutton of the University of Texas at San Antonio. Rabel and Dutton (2014) estimate the capacity of the "massive sand" in the Carrizo Aquifer based on thickness and porosity determined from geophysical logs. Based on their findings, the area of the ASR site for which this pore volume is calculated is 3,022 acres, with a 37 percent average porosity in the injection interval and the average net sandstone thickness of 198 feet in the injection interval. Total pore volume of the main injection zone is estimated to be 233,400 ± 9,000 acre-feet, with a range from 229,700 acre-feet to 266,000 acre-feet, based on use of different log suites (density porosity, total combinable magnetic resonance and sonic logs) (Rabel and Dutton, 2014).

Azobu and Dutton (2014) determined relative volume fractions of Edwards and Carrizo aquifer water within the SAWS ASR site through geochemical modeling. The two waters are geochemically distinct; however, several factors contribute to complexity in estimating the representative fractions. The average Edwards volume fraction is approximately 83 percent based on recovered groundwater samples (Azobu and Dutton, 2014).

Given the importance of the ASR performance to the success of the HCP, appropriate due diligence should be applied to verify the future long-term reliability of the ASR system. Long-term reliability of the ASR system is particularly important if the HCP considers using the ASR system after the current EAA-SAWS contract expires in 2028. Determining the long-term reliability will likely involve additional data collection and studies. The Committee requested information on various water quality, rock-water interaction, and well performance issues. For example, pyrite is reported as a trace mineral in the Carrizo Aquifer (Pearson and White, 1967). This mineral is known to be associated with arsenic and other constituents that may cause water quality issues under variable redox conditions such as ASR (Arthur et al., 2005; Maliva et al., 2006; NRC, 2008). While reports exist for assessing water mixing and water-rock interaction (Otero and Petri, 2010; Malcolm Pirnie, Inc. et al., 2011; Crow, 2012), data are not available to fully evaluate the effects of these processes on stored and recovered water quality over time. The information provided by SAWS did not provide sufficient data to address the potential issues of concern. As water-resource strategies evolve at the Twin Oaks ASR system, any changes in water sources to be mixed with Edwards Aquifer water should be assessed for water quality changes related to mixing of multiple sources, as well as water-rock interactions, to determine potential for adverse water quality changes. Laboratory bench-scale studies, appropriately designed field testing, and geochemical modeling are important tools in this regard.

During a field trip held on February 2, 2016, committee members discussed the performance of the ASR system with SAWS staff. SAWS staff indicated that ASR well injection performance has remained consistent, and they have not observed adverse well performance issues since the ASR began operations in 2004. SAWS staff also indicated they have not observed any adverse rock-water interactions. The EAA and SAWS should give further consideration to evaluating the following items, which may be potential issues in future utilization of the ASR system:

1. Are there any geochemical reactions between the Edwards Aquifer injected/recharged water and the aquifer permeable matrix that may cause adverse water quality issues in the short or long term, especially as the storage volume increases to encounter aquifer matrix not yet exposed to the Edwards Aquifer groundwater?

2. Are there any geochemical reactions between the injected Edwards Aquifer groundwater and native Carrizo Aquifer groundwater that may cause adverse water quality issues in the short or long term?
3. Is there any evidence of mineral precipitation in the aquifer or on well materials (e.g., models or projections of porosity declines in the ASR storage zone) that may affect long-term system performance?
4. What are the long-term trends in ASR well performance? Pyne and David (2005) and the National Groundwater Association (2014) describe the importance of monitoring ASR well performance. A common measure of injection well performance is the injection rate divided by the head rise (specific injection).

The Committee recommends that the following activities be initiated: (1) at a minimum of annually, determine specific injection at each ASR well to assess if there are any long-term changes in ASR well performance, (2) design and implement water quality monitoring for arsenic and related constituents in monitoring wells during recharge and storage events, and (3) design and implement water quality monitoring in ASR wells during recovery events.

CONCLUSIONS AND RECOMMENDATIONS

Implementation of key M&M measures, which are critical to the success of the HCP, is moving in the right direction, with the various programs being characterized by competent project teams, sustained effort, and adequate initial performance monitoring. In general, **for every M&M measure implemented, performance monitoring should be done not only for the first year, but regularly during implementation, with a comprehensive synthesis of the monitoring data about every five years that goes beyond the simple trends analyses found in the HCP annual reports.** The following specific recommendations pertain to individual M&M measures.

SAV Removal and Restoration. Substantial progress has been made removing non-native vegetation from both the Comal and San Marcos systems and replacing it with native SAV species. Nonetheless, despite this sustained effort, **there is not enough new habitat from native plantings to maintain populations of fountain darter to balance non-native SAV removal.** This should be verified by considering the carrying capacity of the various SAV species (both native and non-native) for fountain darter.

Sediment Management. In general, **sediment removal activities should be limited to areas where ongoing upland sources or natural stream dynamics will NOT lead to deposition of new sediment within a matter of years.**

Dissolved Oxygen Management in Landa Lake. **The Committee recommends that aeration not be used routinely as a mitigation measure.** If floating mats cover more than 25 percent of the surface of Landa Lake and dissolved oxygen concentrations decrease, then manual breaking up and removal of the floating mats should be considered as a mitigation measure. Monitoring of dissolved oxygen concentrations using the miniDOTs in selected areas of Landa Lake and Upper Spring Run should be incorporated into an integrated water quality and biological monitoring program.

Voluntary Irrigation Suspension Program Option. **When the HCP is reviewed for renewal, it may be appropriate to re-evaluate the time period that the VISPO trigger is based on using a Decision Support System.** Consideration should be given to redefining the trigger to use additional information, such as groundwater elevation from a longer time frame, precipitation and recharge data, and groundwater model projections of future conditions.

Aquifer Storage and Recovery. **The Committee recommends that the following activities related to aquifer storage and recovery be initiated: (1) at a minimum of annually, determine specific injection at each ASR well to assess if there are any long-term changes in ASR well performance, (2) design and implement water quality monitoring for arsenic and related constituents in monitoring wells during recharge and storage events, and (3) design and implement water quality monitoring in ASR wells during recovery events.**

All Spring Flow Protection Measures. The total expense to implement the HCP in 2015 was $16,397,097 (Blanton and Assoc., 2016), with the spring flow protection measures accounting for 67 percent of total expenses. Due to the high expense of the spring flow protection measures and their importance to the HCP's success, **the Committee recommends that compliance of the parties participating in the spring flow protection measures be audited** so that there is assurance that parties are complying with the terms of the program and the program will operate as designed.

REFERENCES

Arthur, J. D., A. A. Dabous, and J. B. Cowart. 2005. Chapter 24: Water-rock geochemical considerations for aquifer storage and recovery: Florida case studies, *in* Tsang, C-F. and Apps, J.A., eds., Underground Injection Science and Technology, Developments in Water Science. Amsterdam, Elsevier, v. 52, pp. 327-339.

Azobu, J. O., and A. R. Dutton. 2014. Using Geochemical Modeling (PHREEQC) to Determine Edwards Volume Fraction in Mixed Groundwaters from the Edwards and Carrizo Aquifers at San Antonio Water System Aquifer Storage and Recovery (ASR) Site, Bexar County, Texas.

BIO-WEST. 2015. Supplemental Dissolved Oxygen Evaluation in Landa Lake.
BIO-WEST, Center for Reservoir and Aquatic Systems Research, and Aqua Strategies. 2015. Algae and Dissolved Oxygen Dynamics of Landa Lake and the Upper Spring Run.
BIO-WEST, Inc. and Watershed Systems Group, Inc. 2016. Submerged Aquatic Vegetation Analysis and Recommendations. Edwards Aquifer Habitat Conservation Plan Contract No. 15-7-HCP June, 2016.
Blanton and Associates. 2015. Edwards Aquifer Habitat Conservation Plan 2014 Annual Report. Submitted to the U.S. Fish and Wildlife Service March 13, 2015.
Blanton and Associates. 2016. Edwards Aquifer Habitat Conservation Plan 2015 Annual Report. Submitted to the U.S. Fish and Wildlife Service March 22, 2016.
Crow, C. L. 2012. Water-level altitudes and continuous and discrete water quality at and near an aquifer storage and recovery site, Bexar, Atascosa, and Wilson Counties, Texas, June 2004–September 2011: U.S. Geological Survey Scientific Investigations Report 2012–5260, 85 pp., 7 apps.
EAHCP. 2015. January 15, 2015 report of the Regional Water Conservation Program Work Group.
Malcolm Pirnie et al. 2011. An assessment of aquifer storage and recovery in Texas, Texas Water Development Board Report # 0904830940, 200 p. https://www.twdb.texas.gov/innovativewater/asr/projects/pirnie/doc/2011_03_asr_final_rpt.pdf.
Maliva, R. G., W. Guo, and T. M. Missimer. 2006. Aquifer Storage and Recovery: Recent Hydrogeological Advances and System Performance. Water Environment Research 78(13):2428–2435.
National Groundwater Association (NGWA). 2014. Best Suggested Practices for Aquifer Storage and Recovery. Approved by NGWA Board of Directors on July 16, 2014.
National Research Council (NRC). 2008. Prospects for Managed Underground Storage of Recoverable Water. Washington, DC: The National Academies Press.
Otero, C. L., and B. L. Petri. 2010. Quality of groundwater at and near an aquifer storage and recovery site, Bexar, Atascosa, and Wilson Counties, Texas, June 2004–August 2008: U.S. Geological Scientific Investigations Report 2010–5061, 34 pp.
Pearson Jr., F. J., and D. E. White. 1967. Carbon 14 ages and flow rates of water in Carrizo Sand, Atascosa County, Texas. Water Resources Research 3: 251–261.
Pyne, R., and G. David. 2005. Aquifer Storage and Recovery: A Guide to Groundwater Recharge Through Wells. 2nd ed. Gainesville, FL: ASR Press.
Rabel, B. C., and A. R. Dutton. 2014. Estimation of Volumetric Capacity of an Aquifer Storage and Recovery (ASR) Field Operated by San Antonio Water System, Bexar County, Texas.

Acronyms

ASR	aquifer storage and recovery
BioMP	biological monitoring program
BMP	best management practice
CFS	cubic feet per second
CSRB	Comal Springs riffle beetle
DO	dissolved oxygen
DSS	decision support system
EAA	Edwards Aquifer Authority
EAHCP	Edwards Aquifer Habitat Conservation Plan
EARIP	Edwards Aquifer Recovery Implementation Program
ERPA	Environmental Restoration and Protection Area
ESA	Endangered Species Act
FAB	Texas State University Freeman Aquatic Building
FD	fountain darter
FWS	U.S. Fish and Wildlife Service
HCP	Habitat Conservation Plan
HSPF	Hydrological Simulation Program—Fortran
IPMP	Integrated Pest Management Plan

ITP	Incidental Take Permit
LID	Low Impact Development
M&M	minimization and mitigation
MSL	mean sea level
NAS	The National Academy of Sciences
NASEM	The National Academies of Sciences, Engineering, and Medicine
NOAA	National Oceanic and Atmospheric Administration
NRC	National Research Council
O&M	Operation and Maintenance
PAHs	polycyclic aromatic hydrocarbons
PDS	passive diffusion sampling
PDSI	Palmer Drought Severity Index
PPCP	pharmaceutical and personal care products
RBA	rapid bioassessment
RRWG	Recommendations Review Work Group
RWCP	Regional Water Conservation Program
SAV	submersed aquatic vegetation
SAWS	San Antonio Water System
SMARC	U.S. Fish and Wildlife Service San Marcos Aquatic Resource Center
SOP	standard operation procedure
SRP	soluble reactive phosphorus
STR1	Streamflow-Routing Package
SWB	soil-water-balance
SWRI	Southwest Research Institute
TCEQ	Texas Commission on Environmental Quality
TPWD	Texas Parks and Wildlife Department
TWR	Texas wild rice
USGS	U.S. Geological Survey
VISPO	Voluntary Irrigation Suspension Program Option
WQ	water quality
WSTB	Water Science and Technology Board

Appendix A

Evaluation of the Predictive Ecological Model for the Edwards Aquifer Habitat Conservation Plan: An Interim Report as Part of Phase 2

Committee to Review the Edwards Aquifer Habitat Conservation Plan

Water Science and Technology Board

Division on Earth and Life Studies

The National Academies of
SCIENCES · ENGINEERING · MEDICINE

THE NATIONAL ACADEMIES PRESS
Washington, DC
www.nap.edu

THE NATIONAL ACADEMIES PRESS 500 Fifth Street, NW Washington, DC 20001

Support for this study was provided by the Edwards Aquifer Authority under Cooperative Agreement No. 13-666-HCP. Any opinions, findings, conclusions, or recommendations expressed in this publication do not necessarily reflect the views of any organization or agency that provided support for the project.

International Standard Book Number-13: 978-0-309-44489-7
International Standard Book Number-10: 0-309-44489-6
Digital Object Identifier: 10.17226/23577

Additional copies of this report are available for sale from the National Academies Press, 500 Fifth Street, NW, Keck 360, Washington, DC 20001; (800) 624-6242 or (202) 334-3313; http://www.nap.edu.

Copyright 2016 by the National Academy of Sciences. All rights reserved.

Printed in the United States of America

Suggested citation: National Academies of Sciences, Engineering, and Medicine. 2016. *Evaluation of the Predictive Ecological Model for the Edwards Aquifer Habitat Conservation Plan: An Interim Report as Part of Phase 2*. Washington, DC: The National Academies Press. doi: 10.17226.23577.

The National Academies of
SCIENCES · ENGINEERING · MEDICINE

The **National Academy of Sciences** was established in 1863 by an Act of Congress, signed by President Lincoln, as a private, nongovernmental institution to advise the nation on issues related to science and technology. Members are elected by their peers for outstanding contributions to research. Dr. Marcia McNutt is president.

The **National Academy of Engineering** was established in 1964 under the charter of the National Academy of Sciences to bring the practices of engineering to advising the nation. Members are elected by their peers for extraordinary contributions to engineering. Dr. C. D. Mote, Jr., is president.

The **National Academy of Medicine** (formerly the Institute of Medicine) was established in 1970 under the charter of the National Academy of Sciences to advise the nation on medical and health issues. Members are elected by their peers for distinguished contributions to medicine and health. Dr. Victor J. Dzau is president.

The three Academies work together as the **National Academies of Sciences, Engineering, and Medicine** to provide independent, objective analysis and advice to the nation and conduct other activities to solve complex problems and inform public policy decisions. The National Academies also encourage education and research, recognize outstanding contributions to knowledge, and increase public understanding in matters of science, engineering, and medicine.

Learn more about the **National Academies of Sciences, Engineering, and Medicine** at www.national-academies.org.

The National Academies of
SCIENCES • ENGINEERING • MEDICINE

Reports document the evidence-based consensus of an authoring committee of experts. Reports typically include findings, conclusions, and recommendations based on information gathered by the committee and committee deliberations. Reports are peer reviewed and are approved by the National Academies of Sciences, Engineering, and Medicine.

Proceedings chronicle the presentations and discussions at a workshop, symposium, or other convening event. The statements and opinions contained in proceedings are those of the participants and have not been endorsed by other participants, the planning committee, or the National Academies of Sciences, Engineering, and Medicine.

For information about other products and activities of the National Academies, please visit nationalacademies.org/whatwedo.

COMMITTEE TO REVIEW THE EDWARDS AQUIFER HABITAT CONSERVATION PLAN

DANNY D. REIBLE, *Chair*, Texas Tech University, Lubbock
JONATHAN D. ARTHUR, Florida Department of Environmental Protection, Tallahassee
M. ERIC BENBOW, Michigan State University, East Lansing
ROBIN K. CRAIG, University of Utah, Salt Lake City
K. DAVID HAMBRIGHT, University of Oklahoma, Norman
LORA A. HARRIS, University of Maryland Center for Environmental Science, Solomons
TIMOTHY K. KRATZ, University of Wisconsin, Madison
ANDREW J. LONG, U.S. Geological Survey, Tacoma, Washington
JAYANTHA OBEYSEKERA, South Florida Water Management District, West Palm Beach
KENNETH A. ROSE, Louisiana State University, Baton Rouge
LAURA TORAN, Temple University, Philadelphia, Pennsylvania
GREG D. WOODSIDE, Orange County Water District, Fountain Valley, California

Staff

LAURA J. EHLERS, Study Director
BRENDAN McGOVERN, Senior Program Assistant

ACKNOWLEDGMENTS

This report has been reviewed in draft form by individuals chosen for their diverse perspectives and technical expertise. The purpose of this independent review is to provide candid and critical comments that will assist the institution in making its published report as sound as possible and to ensure that the report meets institutional standards for objectivity, evidence, and responsiveness to the study charge. The review comments and draft manuscript remain confidential to protect the integrity of the deliberative process. We wish to thank the following individuals for their review of this report: James J. Anderson, University of Washington; Alan Hastings, University of California, Davis; Jesse C. Jarvis, University of North Carolina, Wilmington; Kenneth A. Moore, Virginia Institute of Marine Sciences; and Steve Railsback, Lang, Railsback & Associates.

Although the reviewers listed above have provided many constructive comments and suggestions, they were not asked to endorse the conclusions or recommendations nor did they see the final draft of the report before its release. The review of this report was overseen by Patrick L. Brezonik, University of Minnesota, and R. Rhodes Trussell, Trussell Technologies, Inc. They were responsible for making certain that an independent examination of this report was carried out in accordance with institutional procedures and that all review comments were carefully considered. Responsibility for the final content of this report rests entirely with the authoring Committee and the institution.

CONTENTS

INTRODUCTION	132
PROGRESS TO DATE	133
MODELING OBJECTIVES AND USAGE	134
MODEL CONFIGURATION	138
MODEL CALIBRATION AND TESTING	150
MODEL COUPLING	154
CONCLUDING REMARKS	156
REFERENCES	157

INTRODUCTION

An ad hoc committee of the National Academies of Sciences, Engineering, and Medicine is in the process of reviewing the many different scientific initiatives under way to support the Edwards Aquifer Habitat Conservation Plan (HCP). The Committee to Review the Edwards Aquifer Habitat Conservation Plan is focusing on the adequacy of information to reliably inform assessments of the HCP's scientific initiatives, ensuring that these initiatives are based on the best-available science. Relationships among proposed conservation measures (including flow protection measures and habitat protection and restoration), biological objectives (defined by the HCP as specified flow rates), and biological goals (such as maintaining populations of the endangered species) are central to the HCP, and are being evaluated during the Academies review. The study spans from 2014 to 2018 and will result in three reports. At the conclusion of Phase 1, the Committee issued its first report (NRC, 2015), which focused on hydrologic modeling, ecological modeling, water quality and biological monitoring, and the Applied Research Program. The Committee will issue its second report in late 2016 and its third and final report in 2018.

This interim report is part of Phase 2 activities and will be incorporated, as an appendix, into the second report. This interim report focuses on the ecological modeling only and is being provided prior to the issuance of the second report in order for the Committee's comments (which take the form of conclusions and recommendations) to be considered while the ecological modeling team is still in place. The final version of the ecological model is scheduled for completion by December 2016, roughly the same time as the issuance of the second committee report. The statement of task for Phase 2 of the Academies study is in Box A-1. This interim report addresses the first item and partially the third item in the statement of task, as they relate to the ecological modeling. At the time of this writing, the ecological models were not developed enough to address Task 3 completely, but such an evaluation will appear in the final Phase 2 report.

This review of the ecological modeling is based on many sources, including presentations made to the Committee and written reports. Presentations documenting the progress of the ecological modeling were given to the Committee in February 2014, May 2014, October 2015, and February 2016. The model development team also provided a report titled "Predictive Ecological Modeling for the Comal and San Marcos Ecosystem Project" (BIO-WEST, 2015) just prior to the February 2016 meeting. Members of the model development team have also made themselves available to answer questions from the Committee outside of committee meetings, including as recently as March 2016. The Committee wants to acknowledge the cooperation and openness of the model development team and

APPENDIX A   *133*

> **BOX A-1**
> **Phase 2 Statement of Task**
>
> The charge to the Academies Committee for the second report states that the Committee will:
>
> 1. Evaluate progress and modifications implemented as a result of the Committee's first report,
> 2. Continue to assess the methods of and data collected through the water quality monitoring and biomonitoring programs,
> 3. Identify those biological and hydrologic questions related to achieving compliance with the HCP's biological goals and objectives that the ecological and hydrologic models should be used to answer, specifically including which scenarios to run in the models. These questions shall help generate information needed to make the HCP Phase 2 strategic decisions about the effectiveness of conservation measures.
> 4. Provide an evaluation of how the Phase 1 conservation measures in the HCP (including flow protection measures and habitat restoration measures) are being implemented and monitored. Specifically, the committee will discuss if the proper method of implementation is being utilized to achieve the maximum benefit to the Covered Species.

the Edwards Aquifer Authority (EAA) to the Committee's questions and inquiries, as this greatly helped the review process.

This review is organized around the four general topics of (1) modeling objectives and usage, (2) model configuration, (3) model calibration and testing, and (4) model coupling. First, a summary of progress to date is presented for the fountain darter (FD) and submersed aquatic vegetation (SAV) modeling. Then, for the first three topics above, the SAV and FD modeling are discussed separately because these topics apply to the FD and SAV modeling as standalone models. The remaining topic on coupling discusses how the SAV and FD models can be developed to enable them to be run so that the SAV model informs (provides inputs to) the FD model. The final section provides a summary and some overarching thoughts about the progress of the ecological modeling.

PROGRESS TO DATE

The modeling effort has made good progress, and scientifically sound frameworks for both the SAV and FD modeling are in place. However, like all ecological and other types (e.g., groundwater) of modeling, the progression through the development, testing, and usage steps of modeling is iterative. Testing often leads to further development as model-data disagreements lead to changes in the model, which is then modified and tested again. Thus, additional effort remains if the ecological model is to be an effective tool for predicting FD and SAV responses to actions that are designed to achieve the HCP's biological goals and objectives.

Trying several alternatives for the SAV modeling was a strategically and scientifically sound decision. Existing SAV models are not designed to address the specific questions of the HCP, and thus trying multiple approaches to the modeling is appropriate to increase the likelihood of success. Using an individual-based approach for the FD modeling was also sound, since such an approach enables more direct and intuitive representation of how spatial and temporal variation in environmental factors important to FD (including flow) will affect FD growth, mortality, reproduction, and movement and the resulting population dynamics. Ecological models like the SAV and FD models can be difficult to fully document, but based on the December 2015 report (BIO-WEST, 2015), the Committee believes that the model development team is on a good track for providing sufficiently detailed description of the models. Although the SAV and FD models are on the right trajectory, it is too early to provide a conclusive statement about the credibility of the models and their eventual usefulness for the HCP-based analyses because many of the details are not fully worked out yet. This is not unexpected, as part of the Committee's approach in this review is to provide input during the development process so it can be considered while the modeling is ongoing.

MODELING OBJECTIVES AND USAGE

The goal of the modeling is clear: develop predictive models to evaluate HCP actions on SAV and FD populations [Section 6.3.3 of the HCP (EARIP, 2012); pages 1-2 of BIO-WEST (2015)]. The objectives to achieve this goal have also been well stated in presentations as well as in the modeling report (BIO-WEST, 2015). In this situation, the developers have a very clear purpose for developing the models, and that seems to be well understood by the development team. Part of the objectives is to use these models in exploratory and scenario analyses designed to assess HCP actions. This would include model simulations designed to quantify how different scenarios of spring flows and conservation measures under the HCP would affect SAV biomass and distribution and consequently FD population dynamics. The issues discussed in the sections below relate to the steps taken and decisions made by the model developers in their quest to achieve these objectives.

General Comments

1. A simple one-time transfer of the models from the developers to the EAA should be avoided because this can result in inefficient, and even possibly erroneous, use of the FD and SAV models.

The knowledge, assumptions, and decisions made by the modeling team during model development are important for subsequently using the models in an effective manner. Also, key questions can be more fully addressed, and additional questions can be addressed, by having the ability to make structural changes in the models, rather than being limited to parameter changes or to a small subset of possible changes determined by the development team before a model hand-off. For example, one can envision wanting to know the range of model predictions to altered flows and thus want to allow for variations of the FD movement rules. It is likely that there will be limited options for the user to change the parameters in the movement rules for FD or say, to try different growth formulations for SAV. Further, the user interface will likely limit the user to select from pre-programmed flow time series options (e.g., different years). It will not be long before the users will want to try other flow time series or variations on drought conditions in order to understand the full range of possible SAV and FD responses. Even if the final version of the model makes many parameters and inputs accessible to the user, there will be interest and demands that require structural changes to the models. One example could be relating FD growth to flow (uncoupled in the present model version), which can be easily done, but requires changing the equations themselves within the NetLogo® code. This might be an easy task for the model developers and other experienced modelers, but could be perceived as off limits (thereby limiting the usefulness of the models) or easily done incorrectly by less experienced people unfamiliar with the code.

The situation with these ecological models is the same as with the groundwater, hydraulic, and other models being used by EAA: the FD and SAV models are iteratively improved, and their use requires certain ongoing expertise to be available. A small pool of people is needed to curate the FD and SAV models to ensure they are used effectively and with the proper flexibility to allow examination of questions and incorporation of new data. This pool can involve in-house expertise as well as access to the model developers.

Models such as the SAV and FD models can generate variables (e.g., FD abundances) that can be over-interpreted, such that some caution and "management of expectations" is needed to ensure that the models are used to address appropriate questions and the results are properly interpreted. Factors such as flow can be explicitly or implicitly included in ecological models; both allow for investigation of how changes in the factor affect

FD or SAV, but both also require careful evaluation of how the factor is actually used within the model. For example, if an input is labelled "flow," simply changing its value and interpreting the results may not reveal what would happen if flow changed in the actual system. Similarly, if there is no input labelled "flow" this does not mean flow effects cannot be examined; for example, changing vegetation coverage in the FD model implies some impact of flow, even if flow is not explicitly included. In this way changing the available inputs that are affected by flow (the implicit approach) allows for scenarios of changed flow regimes to be evaluated. Managed expectations also apply to model output. It is unlikely that simulation runs, while spatially explicit, will provide maps that will directly mimic the actual environment. Interpreting the model output is sometimes appropriate as native units (abundance of FD; biomass of SAV) and in other situations should be interpreted as changes in abundance or biomass from a baseline simulation (a percent change).

Fountain Darter

2. The focus on using the FD model to predict the responses of FD abundance to alternative HCP flow control packages is useful, but there are other uses of such mechanistic models that should be considered.

Two of the most powerful uses of the FD model beyond predicting total abundance are to (1) provide a systematic analysis of what life stages, processes, timing, and spatial areas are important to FD population dynamics; and (2) include explanations of why model responses are predicted (not just the final predictions of abundance). The plan for model usage seems to underplay these uses. The idea of running different flow time series through the model is a good starting point, but stopping there would not utilize the full benefits of having this type of model. Also informative would be to tease apart what aspects of the HCP flows cause the population responses, such as simulating the FD response to synthetic flow time series that systematically vary the pattern, peaks, and troughs of the flows. In addition, all key simulations should be accompanied with explanations as to *why* the population responses occurred within the model. What changed in FD growth, mortality, reproduction, and movement, by life stage, between the simulations that used two different flow time series? A convenient way to summarize the relatively complicated output of individual-based models is to use life tables and to estimate summary measures from the life tables such as the finite population growth rate (λ) for that year (λ values are reported for some FD model results already), and to perform follow-up simulations that specifically vary what was identified as key changes but to do so in an experimental design. Suppose the altered HCP flows resulted in a 15 percent increase in the FD population adult abundance over the 10 years of the simulation. One should then use the outputs and additional simulations to identify what aspects of the altered flows (e.g., a particular year or sequence of years), and which FD processes and life stages, contributed to the increased population abundance.

A model such as the FD model can also be used very effectively in an "inverse" mode. Simulations can be performed to identify which processes and life stages are sensitive to changes in flows, and then these viewed seasonally and spatially to see how they match up with HCP actions. Actions that affect highly sensitive processes and stages can be considered, at least in the virtual world, of having a higher likelihood of impact at the population level. Limiting model usage to simulating population abundance trajectories for flows with and without HCP actions would underutilize the management potential of the FD model.

Submersed Aquatic Vegetation

3. The goal of creating an SAV model that simulates dispersal and predicts how flow affects SAV has not yet been met.

The objectives of the SAV modeling are to predict the percent SAV coverage under different flow regimes, and to then use these predictions as input (habitat) to the FD model. However, mechanisms connecting flow to SAV coverage in the SAV model are presently limited to how changing depth (as a result of flow) affects light availability. Unlike the FD model, the SAV model is a mass-based model (not individual-based), with both implemented on the same spatial grid. At this point in model development, the focus in the development of the SAV model has been on how light impacts SAV biomass. Indeed, if light availability is the single forcing under consideration, a simple cellular automata approach may yield similar predictions with a lower level of complexity. A fundamental shift in emphasis to understanding how flow affects SAV seems to better align with the objectives of how this model will interface with the FD model and with assessing the ecological responses to HCP actions.

Developing rules to approximate dispersal within SAV models is an ongoing challenge in the SAV modeling field. The difficulty is specifying sufficient rules that incorporate dependence of dispersal on the appropriate environmental and biological factors and result in SAV composition and biomasses that realistically change in time and space. Here, with multiple species being simulated, the potential for considering plasticity in the responses to flow is also compelling. The model development team is currently grappling with the challenges of modeling dispersal, and the interim report includes some compelling and creative ideas for simulating this process. Spending time considering how flow affects these processes is critical. It will also be important for the modelers to carefully consider how the dispersal model interfaces with the biomass-growth model for SAV. As described, the dispersal model could easily be considered as a separate modeling exercise, but its effectiveness will be improved by ensuring that it is appropriately matched to the approach for simulating SAV growth.

Certainly expanding on the plans to incorporate scouring impacts seems valuable, including very low-flow impacts in the lake systems (increased epiphytes or temperature). Every decision in these models should be carefully examined against the overarching question regarding how flow affects SAV, and in this way additional processes will be identified that connect flow to SAV dynamics. Essentially, the processes included and the characteristics of the model formulations serve as hypotheses regarding how the ecologists working in this system might consider the impacts of flow on SAV. The strength of modeling is that many of these hypotheses may be evaluated in a simulation setting as a first cut to determine whether they are critical to understanding the impacts of flow on SAV.

MODEL CONFIGURATION

Model configuration includes the specification of the spatial and temporal scales of the model, the state variables to be tracked, and what processes are included and how they are represented.

General Comments

4. The temporal and spatial scales of the SAV and FD models are reasonable but the representativeness of selected reaches and the variance properties associated with the use of QUAL2E outputs as model inputs should be clearly documented.

The temporal and spatial scales of the FD and SAV models should be defined based on the key aspects of the driving variables (e.g., flows), the rates of the processes to be simulated, and the questions to be addressed. In addition, the temporal and spatial scales need to be compatible. For the FD model, an hourly time step and 1 m^2 cells are reasonable decisions, although the spatial resolution seems relatively fine compared to the time step. Fish trying to forage, avoid predators, or prevent localized overcrowding can move potentially many cells in one hour. The movement algorithm needs to be capable of dealing with realistic distances moved in a time step. For the SAV model, daily time steps and 1 m^2 cells are reasonable, although permitting colonization only once per month may not capture lateral growth of these clonal plants. Using daily averaged values of flow as a forcing for the SAV model is likely adequate for simulating depth and light availability, but may not permit incorporation of additional mechanisms related to flow, such as uprooting or dispersal.

Model inputs include the hydraulics and water quality outputs from the QUAL2E model, with the FD model also receiving inputs from the SAV modeling. Collapsing the resolution of the two-dimensional (2-D) grid of QUAL2E from 0.25 m^2 to 1 m^2 cells was a reasonable decision by the development team; care should be taken in how the predictions of the QUAL2E are aggregated. The QUAL2E modeling also has a fast time step so its results can be summarized to match the hourly time step of the FD. Whenever aggregations are done, it is advisable to keep track of the loss of variance in the transferred variables (e.g., hourly variations around a daily average flow; value of four cells to one value for the larger cell) and whether different aggregation schemes (snapshot versus averaging versus daily minimum) affect the values of the transferred variables.

The spatial domain of the FD model is not simply the area that encompasses the number of FD individuals (abundance) expected in their entire geographic range. Rather, the FD model simulates individuals in certain reaches (subregions) of the system affected by the HCP. How well these subregions, simulated independently, represent the area inhabited by the entire FD population should be confirmed. (This issue of the representativeness of regions was discussed extensively in Chapter 4 of NRC, 2015.) For the SAV model, simulations at the reach scale are useful for predictions of HCP-related effects and also for model validation purposes.

Fountain Darter

5. The use of an individual-based approach imbedded within a 2-D spatial grid for full life-cycle simulations of FD population dynamics is a scientifically sound framework for the questions being asked, but there remain some important steps to link the FD dynamics to their habitat.

The parallel development of the FD and SAV modeling has advantages in that adjustments can be made in each to ensure both models are configured to allow accurate transfer of habitat information from the SAV to the FD models. It is planned that the FD model will require the output of the SAV model, but the SAV modeling is not affected by the dynamics of FD. Currently, the FD model is not using results of the SAV modeling as inputs of habitat; rather the FD model is using inputted field data-derived habitat maps that abruptly update every six months (uncoupled mode). This is a reasonable temporary fix in order for the development of the FD model to continue while the SAV modeling gets refined. However, because the uncoupled approach uses observed SAV maps, habitat in the FD model is not directly linked to flow. Therefore, the uncoupled version, in its present form, cannot be used to examine HCP-related scenarios involving changes in flow. The coupling of the SAV and FD models is discussed below.

6. The representation of the processes of FD growth, mortality, reproduction, and movement presently in the model are well-founded but may be too simple and not sufficiently linked to changes in habitat and flow to answer some of the important management questions.

Growth is presently represented as fixed in the FD model. That is, stage durations determine the progression from one life stage to the next, and these durations do not vary within or between simulations. Thus, the approach implicitly includes growth rate of individuals, but body length and weight are not tracked as state variables. Sometimes this approach is misinterpreted as assuming that food is not limiting. The degree of food limitation is determined by how the durations are estimated; if estimated from the field and food was limiting in the field conditions, then the durations reflect highly averaged but still food-limited conditions. However, the fixed-stage duration approach does make the strong assumption that the availability of food does not vary much from the conditions under which the durations were determined. The present version of the FD model assumes that individuals will obtain the food needed to achieve the growth rates dictated by the durations, and these growth rates do not vary much in space, seasonally, based on the specific habitat being inhabited, or based on flow. Thus, the ability for growth of individual FD in the model to respond to variation in environmental and habitat conditions, including HCP-related actions, is very limited. The biological realism of this limitation, and how it affects the usefulness of the model, should be evaluated.

Mortality is represented as stage-specific rates plus additional rates dependent on temperature and movement. The movement-related mortality rate is triggered when the number of movement time steps (24 per day) that an individual spends in open water or without options to move to other less crowded vegetated cells is exceeded (see Movement paragraph below). When an individual dies, it is removed from the simulation. This representation of mortality related to movement being density-dependent is critical because it is the only source of density-dependent control on the FD population within the model. It only operates at relatively high FD abundances (so no depensatory mortality is represented), and it only occurs when SAV habitat is limited relative to FD densities. The role of flow is, at best, an indirect effect through flow affecting SAV; however, such dependence of SAV on flow is not presently in the FD model.

Reproduction is relatively fixed in the FD model, with maturity dictated by the fixed stage durations until the adult stage and fecundity fixed at

19 eggs per batch per female. The aspect of reproduction that can vary is based on vegetated cells. This is because, if a female is attempting to spawn, the individual must be in a vegetated cell and must not have spawned for at least a month. When these two conditions hold, there are fixed probabilities by month that the individual will spawn and release 19 eggs. Eggs remain in the cell into which they were released as they progress to larvae and then to juveniles; juveniles and adults can move. Reproduction has the potential to be related to habitat and to be density-dependent. For example, if SAV is severely limiting as habitat for FD, then female individuals that could spawn based on the other constraints may not spawn because of the limited availability of vegetated cells. It is not clear how this would occur in the model (e.g., would individuals move to vegetated cells for reproduction?) and whether such severely limiting habitat conditions are realistic.

Movement is a rule-based neighborhood search approach, and it is only triggered under locally crowded conditions. NetLogo® follows individuals in continuous space, and after an individual moves and its position is updated to its new continuous location, the cell that the individual is located in is then determined. The cell location determines the environmental conditions an individual will experience for the next time step. The present version uses a cell-by-cell movement rather than using conditions to determine the x and y velocities of individuals and then updating their continuous locations. The present movement algorithm also uses up to 24 evaluations in a day, which can be confused as being hourly. However, this is not the case because conditions affecting movement do not change hourly but rather change daily (depth, velocity, temperature) or seasonally (vegetation type). The time-stepping of movement within the day is to deal with individuals moving for a day among very small cells (1 m^2) and to allow some exploration by the individual of the local area. An alternative would be to update movement only once per day but to allow an individual to "see" a larger neighborhood than one cell in the four (or eight) directions.

The movement rules are driven by maximum FD densities that are assigned to the vegetation types for each cell that then change seasonally. Movement is triggered when the FD densities in a cell exceed the maximum densities. Some movement between adjacent cells, even if the present cell is not too crowded, is included: if an adjacent cell is also less than maximum density, then there is a 50/50 chance to move there or stay in the presently occupied cell. In the other case of overcrowding in a cell, the individual attempts to move to a neighboring vegetated cell and only can if that cell is not crowded. If all vegetated adjacent cells are also crowded, then the individual would move to an adjacent water cell if there are any. The number of times the individual is in water cells is accumulated and used to determine death (too many time steps in water cells leads to death). An individual can also die if no uncrowded or water cells are available to move into for enough time steps.

Use of a rule-based movement implemented on a cellular (cell to cell moves) scale can realistically represent movement. The difficulties arise when the temporal and spatial scales are not well matched. The approach taken with the FD model to address this potential issue of a coarse (daily) time step with a fine (1-m^2) spatial resolution is to allow for 24 moves within each day. Information on the typical distances moved by individuals and plotting of the Lagrangian trajectories of individuals under different vegetation and flow conditions should be presented to confirm the realism of the simulated movement behavior. Another potential difficulty with a cellular approach to movement is if the spatial resolution of the FD grid is changed—movement to a cell now involves traveling a different distance in the same time step. Finally, there is always debate with a neighborhood search algorithm about what do the individual fish sense and how do they know how to go a neighboring cell without having visited it. The fine spatial resolution of the FD model helps in this case because it is easier to envision individuals detecting gradients and other cues on a 1-m^2 basis that would allow them to "sense" the conditions of the destination cell in advance of moving there.

The only linkage among the growth, mortality, reproduction, and movement processes is how movement can contribute to mortality. This may be reasonable for FD and the questions being asked, but it is very important for the audience to understand this so the results can be properly interpreted and the model used appropriately. Growth is fixed and based on specified durations of life stages; no matter what conditions are simulated, the individuals will always grow at the same rates and progress through the life stages at the same rates. Mortality does not depend on size but only on stage and temperature. Reproduction, which like mortality is often represented as size-dependent in fish population models, is completely size-independent in the FD model. Maturity depends on stage, which depends on growth, which is fixed; fecundity is also fixed per individual. For these reasons, interpretations of modeling results such as "flow caused slower growth and this led to higher mortality and lower reproduction" are impossible. The point is that interpretation of model results and the types of scenarios that can be simulated depend on the structure of the model. In the FD model, few of the possible linkages (see Rose et al., 2001) between growth, mortality, and reproduction are represented. This may be appropriate—it depends on the biology of the species—but is atypical of many individual-based and population models of fish and requires careful consideration as modeling results are reported and interpreted.

7. Thresholds in process representations should be used cautiously because they can erroneously create non-linear population responses and unrealistic sensitivities to changes in habitat and flow.

APPENDIX A

The use of daily maximum and minimum values from QUAL2E as inputs to the FD model should be done carefully. If processes are formulated to depend on maximum or minimum daily values (e.g., minimum dissolved oxygen [DO] affects daily mortality), then the model is internally consistent. However, such formulations should be done cautiously, especially with the relatively smooth changing hourly values of the rest of the processes in the model. One of the advantages of the individual-based approach is that it allows accumulation of hourly exposure of individuals to environmental conditions over time. While using minimum or maximum daily values for each day to affect processes is mathematically valid, formulating how these minimum and maximum values affect processes, which themselves could be a threshold response (rates change suddenly not smoothly), is challenging. *At a minimum, a thorough sensitivity analysis to evaluate the impact of these thresholds seems warranted.* The link from flow to temperature and DO is important because these indirect effects of flow are the only effect of flow on FD to date in the FD model. Thus, interpreting how alternative flows affect FD using the FD model requires understanding how changes in flow affect velocities and depth that are then used as input to the QUAL2E model, and then how these changes in hydraulic outputs affect QUAL2E's predictions of maximum daily temperature and minimum daily DO.

The use of observed densities for maximum FD densities by vegetation type acts to smooth over the threshold effect of capping FD densities by vegetation type. The smoothing occurs because a range of "maximum" densities are used for each vegetation type rather than a single value. A possible inconsistency occurs because observed densities are not truly maximum densities. Nonetheless, the use of observed densities for maximum densities will help in calibration; that is, as SAV types change in the FD model, the maximum densities change, which in turn encourages the model-predicted densities to mimic the observed densities. Total abundance of FD is the sum of their densities over all cells; thus, model-predicted abundance is a direct result of what values the maximum densities are set to. Because the observed densities were used to limit the model and then the calibration and validation use the sum of the simulated densities compared to the sum of the observed densities, the calibration and validation results showing good agreement are not as rigorous as they may seem based on the predicted versus observed abundances plots. This calibration strategy requires some skill because exceeding the specified maximum densities triggers movement, which can result in higher mortality. Proper interpretation of the calibration and validation results is critical for associating the appropriate level of confidence with model predictions of HCP effects.

8. The representation of density-dependence and how its effects on individuals manifest at the population level needs further evaluation.

Density-dependence is when the rates of a process (e.g., mortality) depend on the number of individuals present in a specified area (e.g., particular cell). Density-dependence can occur with growth, mortality, reproduction, and movement (Rose et al., 2001). As with other effects (e.g., flow), density-dependent effects on mortality and reproduction directly affect the number of individuals in the population (abundance). Density-dependent growth and movement are important because they can have indirect effects on mortality or reproduction (e.g., mortality rate decreasing with size); otherwise, changes in growth or movement do not affect abundance. Including density-dependence in population models is important because most density-dependent effects are a negative feedback and act as compensatory mechanisms. They will offset some of the response of the population to changes in habitat and other factors. For example, a decrease in spring flow can cause reduced SAV habitat for FD and increases their mortality rate because of less cover resulting in increased predation. However, the reduction can then be offset to some extent by reduced crowding at spawning, resulting in females releasing more eggs and these having higher survival. Thus, even with fewer spawners, the higher individual fecundity and higher egg survival results in an increased total egg production. (Note: such a logic chain of responses is not possible in the current version of the FD model.) In subsequent years, the reduction in the population is less than what would be expected from the reduced habitat alone under density-independence. Similarly, augmenting habitat would result in less positive response than expected under density-independence. Without density-dependence (no negative feedbacks), populations cannot be stable for extended periods of time because slight changes in reproduction or mortality must result in them either going extinct or growing unbounded.

The representation of density-dependence in the FD model is limited and restricted to increased mortality under relatively extreme local crowding. Each cell is assigned a habitat type, and a maximum density is generated from field data on densities. Increased mortality occurs when movement options are limited to neighboring cells that are also at their capacity. While this triggering of density-dependence when certain crowding conditions occur is a reasonable representation, it is quite limited in scope. There are other aspects of mortality, as well as growth and reproduction, which could be density-dependent. A simple approach that would allow rapid exploration of the importance of density dependence would be to assume that survival, growth, or fecundity decreases a reasonable amount (similar to the range exhibited in data) as density goes up (depending on vegetation type). Simulations with various combinations of the possible density-dependent processes could be analyzed to determine if further effort to refine the relationships is warranted. In general, a clear rationale for what processes are density-dependent—based on the data, expert opinion, and other similar species—should be developed.

However density-dependent is represented, when all effects are simulated on individuals it is important to show how these effects add up to density-dependence mortality at the population-level. For example, a typical diagnostic to use is showing the annual spawner-recruit plot that results from multiple years of simulation. A common measure of spawners is total eggs produced in a year, and a common measure of recruitment would be the number of individuals that survived from those eggs to become juveniles and then to become adults. This can be difficult with a species like FD that spawns all year long and for which the present formulation includes density-dependence in the adult stage; defining over what months to sum egg production and how to accumulate recruits to obtain annual values needs to be considered. In addition, because of its potential importance on population dynamics, the density-dependence in adults should also be characterized and quantified. Based on the life history of the FD, one would expect a Beverton-Holt type spawner-recruit relationship, likely with a weak response (gradually leveling off curve, Figure A-1B). One often characterizes these curves with the steepness coefficient that summarizes the strength of the density-dependence in the spawner-recruit relationship, which has been reported for hundreds of fish species (e.g., Rose et al., 2001). Based on the Committee's experience, a steepness value of 0.5 to 0.7 is anticipated. One could also try to create a spawner-recruit curve using proxies from the field data and compare its properties (e.g., shape) to the model predictions. Some additional exploration of how density-dependence manifests itself at the population-level is needed.

9. The representation of flow effects in the model seems too limited in potential effects due to reliance on having site-specific empirical evidence for the effects.

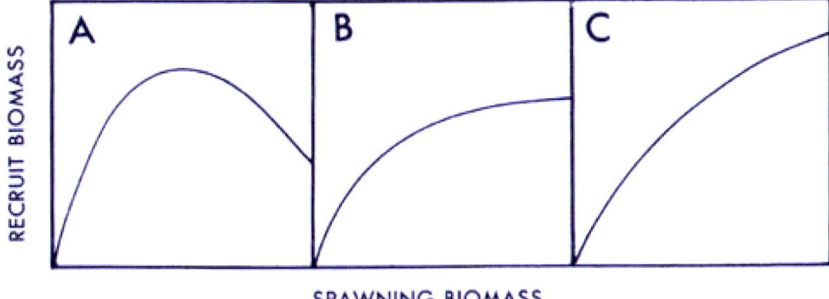

FIGURE A-1 Three common spawner–recruit relationships: (A) Ricker, (B) Beverton-Holt, (C) Cushing. SOURCE: Figure 10 from Parrish and MacCall (1978).

A logic flowchart showing how a change in spring flow affects FD directly and indirectly would be helpful. It must start with flow and eventually result in affecting mortality or reproduction, as these are the two processes that determine FD abundance. Flow effects on growth or movement must then continue in their logic to see how these flow-related changes affect mortality or reproduction. For example, if lower flow affects water depth in cells and this causes FD to move to other cells but their growth, mortality, and reproduction are the same in the new cells, then the lowered flow had no effect on FD abundance despite movement being density-dependent. Similarly, if lower flow was represented as affecting growth rate (i.e., longer or shorter stage durations), this also would have no effect on FD abundance unless mortality rate also was specified as dependent on stage duration. In the present model, mortality rate decreases with stage, and thus prolonged duration in early life stages, with their high mortality rate, could result in higher cumulative mortality. Slowed growth could also result in delayed maturation (reaching the adult stage) and reduced fecundity, but these may or may not have ecologically meaningful effects on population. The logic becomes complicated; does flow affect temperature which then affects mortality or does flow affect SAV, which affects FD habitat? A logic flowchart would enable easier tracking of the direct and indirect effects of changes in flow or other variables affected by the HCP.

With the present configurations of the SAV and FD models, the direct and indirect effects of flow on FD seem to be limited. The direct effects are limited to how flow affects daily maximum temperature and minimum daily DO (from QUAL2E), both of which affect mortality rates. Flow can also indirectly affect FD through flow effects on SAV dynamics, which determines the maximum FD densities in cells, which could lead to movement that causes increased mortality rates. In the uncoupled mode, the observed spatial maps of SAV reflect the effects of flow, but flow is not available to be adjusted in any systematic way (i.e., there is no flow input variable to the SAV maps). When the SAV model is further along in development and the coupled mode is implemented, any indirect effects of flow on FD through SAV will depend on how flow affects the SAV. Present plans, which are subject to adjustment and change as the SAV modeling proceeds, suggest flow could affect the biomass of an SAV species in a cell by altering water depth, which determines light limitation of photosynthesis and temperature affecting respiration. The report also lists velocity directly affecting SAV, but its role it not yet clear. It also has been proposed that the way an SAV species is assigned to a cell (transition) every three months, and maybe also dispersal, could depend on flow, although these remain ideas at this point.

Model development can proceed using several different philosophies, and the approach seemingly taken for the FD model may have over-restricted how flow effects are represented. One philosophy ("top-down") is to focus on formulating the model so that there are relatively few parameters that can then be optimized based on simulated and observed population-level variables (e.g., adult abundance over time). Here the fit between predicted and observed values is critical, and the idea is to avoid over-specification of the model. Another philosophy ("bottom-up") is to carefully develop each component of the model so that when they are put together there is high confidence in the simulated population-level dynamics. The present version of the FD model relies on there being strong empirical evidence for flow effects in order for those effects to be included. In very well studied systems, this is effective because the major possible effects usually have been studied and their representations have a sound empirical basis. However, this approach can lead to over-simplified representations of the effects where the empirical evidence is not strong enough to justify including many of the possible effects that are suspected (e.g., intuitive, data suggestive, occur in other systems) but not documented. Thus, uncertainty due to the lack of site-specific data leads to ignoring possibly important effects. While this system is well-studied in some respects (sampling of FD densities; observational data), many would consider it under-studied in terms of process studies, especially those that relate flow to growth, mortality, reproduction, and movement of FD by life stage. Thus, the FD model reflects what is clearly known about flow effects but likely is missing other effects because of lack of site-specific measurements to justify their inclusion in the model.

There are several approaches for dealing with the possibility of under-studied effects not being considered in models. An excellent use of the FD model would be to add some of these suspected effects and explore how including them would affect model results. One approach is to use information from similar species and other systems to infer, in this case, possible flow effects on growth, mortality, reproduction, and movement. These can be put into a category that distinguishes them from the effects documented using site-specific data so people know there is higher uncertainty (less site-specific evidence) with these effects. One can then use a series of simulations (like a sensitivity analysis) to see if these less-well-known effects could have significant population-level effects and have an impact on the advice provided to management. This use of the FD model also then leads to the identification of uncertain information that is also critical to accurate predictions and how to design sampling or experiments to provide this information on a site-specific basis for later incorporation into the FD model.

Submersed Aquatic Vegetation

10. Use of highly simplified formulations describing nutrient limitation or effects of temperature on photosynthesis may be problematic when the model is applied to scenarios where these factors are critical.

Model development must necessarily simplify the system. Nonetheless, it is critical to document and justify what assumptions and decisions have been made regarding which mechanisms to include or focus on. This justification should explain why certain factors or processes were included and why they were formulated at the level of detail used, as well as state why some factors and processes were not included. To develop an SAV model without considering the impacts of nutrients, as this model does, is highly unusual. It was the recommendation of the Committee's first report (NRC, 2015) that nutrients be measured regularly. Nutrients can be limiting to plant growth and can also result in impaired growth conditions. At low flow conditions, especially in the lake systems that can act as refuges, there could be a future scenario where nutrient issues may be critical. For example, abundant nutrients under low flow conditions may encourage growth of epiphytes that then limit light availability to the SAV. For a model such as this, which is being developed largely to help predict the response of the system to hydraulic conditions not regularly experienced, it seems critical to systematically evaluate the basic factors involved in the growth of the SAV for potential inclusion in the model, level of detail of representation if included, and possible mechanisms linking them to flow.

The treatment of temperature in the model is inconsistent in that there is no temperature limitation in the photosynthesis formulation, but temperature effects are included in respiration and growth equations. Including a temperature limitation term for photosynthesis would resolve these inconsistencies. In many instances, respiration and photosynthesis respond differently to temperature changes, and explicitly including temperature dependencies may be illuminating.

11. In general, more model detail in the final report is critical for both review and future users of the SAV model.

The SAV modeling group has been very helpful in answering questions related to the BIO-WEST (2015) report. Nonetheless, future reports should provide more detail on decisions and assumptions, choices for parameterization, and occasionally referencing of the other coupled models (FD and water quality) in order to aid future users and developers of the SAV (and FD) models. For example, providing greater clarity on the conversions from grams dry weight to glucose and back again, and detailing differences in these conversions amongst species, is important. The Committee's reading of the BIO-WEST (2015) report suggests that light attenuation data are lacking, such that gathering some field data for solar irradiance and light attenuation would improve upon current forcings and fixed parameterization of the k value (light extinction coefficient). The BIO-WEST (2015) report also suggests that basic temperature limitation studies are not in abundant

APPENDIX A

supply for the varied species modeled here and that the impact of temperature on mortality is not strongly understood. Providing referenced literature on these links (e.g., between mortality and temperature) is recommended, as is providing more detail on the relationship between flow and scour. Finally, details on model initialization should be included in the final model description. It would be most effective for the modelers themselves to provide an explicit list of the assumptions made, perhaps in some prioritized list, to aid in future iterations and improvements to the model. The developers have the clearest picture of what data, research, and questions must be pursued to improve future management of the systems and to aid in improvement of the models. Strongly identifying those areas where assumptions were made or data were lacking is an invaluable practice.

12. Many parameters appear calibrated, and it is not clear how the values of fixed parameters are connected to literature values. Formulations are taken from a crop model, which is not a problem as long as the developers sufficiently incorporate SAV morphology, growth, and physiology in the formulations and parameterization. Describing how the calibration is done and convincing end users that the parameterization is appropriately matched to reasonable values from the empirical literature will aid model credibility.

Calibration allows for changes in model parameters until predicted and observed values appear consistent with each other. However, calibration must also include documentation that the tuned parameters are realistic and, wherever possible, match literature and site-specific values. There is little information provided regarding parameterization in the BIO-WEST (2015) report. After some evaluation, the modeling team decided to develop a new model, based on a suite of existing models. The basic growth formulations are borrowed from Teh (2006), which focused on crop models. Using growth formulations from other plants is a common approach used by modelers and is effective and efficient as long as the formulations are carefully checked and adjusted based on SAV information and site-specific information. The model is likely extremely sensitive to P_{max} (the maximum photosynthetic rate for a species), as are most models of this type, such that a thorough sensitivity analysis is advisable. Based on the available documentation, it appears that relatively little data or empirical parameterization has been taken from the SAV literature; a majority of the parameters are calibrated. A rigorous review of the final SAV model will seek examples where these selections are well founded in the SAV literature.

13. The SAV model grapples with the difficult challenge of handling maximum biomass per cell, as well as conversion of cells from one SAV species to another.

Simulating the processes of both colonization and conversion from one species to another is perhaps the most exciting and challenging aspect of the SAV model. The user-determined maximum aboveground biomass is set to limit biomass in a given cell. However, the model already includes self-shading and permits for negative growth as the main mechanisms that should, presumably, impose a more mechanistically derived limit on the maximum amount of biomass in a cell. Forcing a maximum biomass value can artificially help calibration because it simply cuts off biomass values that are too high without a biological reason. If the development team instead considers this limit to be related to colonization of adjacent cells, then this could be explicitly linked to the transition probabilities. Another option would be to have a variable translocation term, where more growth is allocated below ground as the above ground biomass in a cell becomes larger. For the species that float across the surface, thinking through whether it is necessary to have a rule that limits height to the water depth is advisable. Light is simulated with some detail throughout the water column, which may be critical for the range of species simulated in this model, some of which grow basally and some apically. However, it is not clear if this detailed water column light approach is matched with an equally detailed approach to modeling the SAV that takes into account the location of the meristem. Perhaps most importantly, the need for a translocation term suggests that further work may be needed on the rate process portions of the model.

The dispersal model is still under development and generally appears to be sound. Considering a cost to the parent biomass after dispersal to an adjacent cell seems like a reasonable adjustment that may be useful. Describing in greater detail whether the modelers consider the dispersal process to be related to sexual or asexual reproduction could also be helpful. Recognizing that the transition probabilities are currently under development, it is still important to provide more detail as to how they will be coupled to flow and the biomass dynamics portion of the model.

MODEL CALIBRATION AND TESTING

Model testing is the estimation of model parameters (calibration) and testing (validation) of the model's performance. The credibility of the modeling results depends, in large part, on how well the model can generate realistic behavior.

Fountain Darter

14. Calibration and validation of the FD model to date show the model can reproduce the historical abundances, but additional confidence is needed to most effectively use the model for management purposes.

The strategy for calibration of the FD model was to vary the number of movement time steps needed to trigger mortality within the movement rules until simulated population abundance stayed near the maximum possible densities for 2003 to 2014. Additional simulations showed what happens if the density-dependent mortality is relaxed (see almost exponential increase) and if movement was simply random (extinction). The calibration results are reassuring, as they confirm the types of model behavior we expect and they demonstrate that the model can show other types of behavior if not properly constrained, but they could be more convincing. The follow-up validation analysis used the same approach, but with the model applied to other reaches than to which it was calibrated. Thus, the calibration and validation are based on this same strategy of the degree of agreement of simulated abundance hovering around the specified maximum densities over time.

It should be noted that the good agreement between predicted and observed abundances within the calibration is somewhat tautological. This is because, in the model, FD densities are constrained to be less than the maximum densities in each cell; overcrowding kills them if they cannot move to a cell where there is room for more individuals. The maximum densities were set to observed densities by vegetation type. So the fact that the sum of FD densities (abundance) hovers near the sum of the maximum densities (abundance) is somewhat expected if the model was generating roughly realistic densities with some surplus production. This calibration approach would fail (e.g., predict extinction) if mortality was too high or reproduction was too low; the population would decrease, and there is nothing in the model that triggers density-dependence (lowering of mortality or increasing in reproduction) at low densities. If the mortality and reproduction rates were set so that there is sufficient potential to produce adults in the model (e.g., reproduction greater than mortality), then the calibration approach used could be successful. The simulated abundance would try to exceed the specified maximum densities, which would trigger density-dependent mortality (i.e., higher mortality), and the simulated abundances would then decrease; with adjustment of the degree of density-dependent mortality, the simulated abundances would then hover near the summed maximum densities. Based on the calibration results to date, the Committee would characterize the model as being a good descriptor of FD abundances during the period of simulation, rather than being a tool for true prediction or forecasting. This is known by the model developers but it needs to be clearly understood by the general audience. The plots of simulated and maximum densities can mistakenly be interpreted as true model predictions that greatly agree with the maximum densities, which may (wrongly) lead to thinking the model is an excellent independent predictor of absolute abundance or can be used to forecast the response of abundance to large changes in flow. The model may indeed have such capabilities, but the calibration and validation done to date cannot be used to conclude that.

The calibration and validation can be strengthened by examining additional model outputs and years, and by quantifying the uncertainty associated with predictions. Some of this has been done by the model development team but could be better documented, more rigorously compared to the field and lab data, and additional outputs considered. For example, one could examine the simulated spatial distributions and movement trajectories of individual FDs in the model, and perform more in-depth contrasting of dynamics between years with extreme conditions. Scenarios can also be simulated that manipulate flow and SAV (habitat) conditions to then track how these progress through the FD processes and life stages, resulting in population-level responses. The presently used series of years can be manipulated to increase the interannual variation in environmental conditions. Some years can be adjusted or new single or a few years (e.g., drought, scour) inserted. Model responses at selected steps in this changed flow leading to a population response can be qualitatively compared to lab results and field data to confirm such intermediate effects are realistic. Propagating uncertainty and stochasticity through the FD model, while not adding to the validation credibility, would help in ensuring proper interpretation of model results and model differences predicted under different HCP scenarios.

Sensitivity analysis (model response to small changes in inputs) and uncertainty analysis (model response to realistic variations in inputs) can be used to identify key model inputs and the associated variability in model predictions. If key inputs such as parameter values can be identified, then field and lab studies can be designed to provide more certain estimates of these inputs. These revised estimates can then be inserted back into the models to reduce the uncertainty of the predictions. Furthermore, it is important to present not just individual values as model predictions but also the variability around those values. This aids in the comparison of model predictions to field data, as both have variances. Presenting the variability around predictions is also important to properly interpreting the results from running alternative management scenarios—that is, do these scenarios really lead to differences that go beyond the known levels of uncertainty.

15. The historical time period used for calibration had relatively similar environmental conditions from year-to-year, which limits the range of conditions of scenarios feasible for exploration by the model.

The 12 years used for calibration included a relatively narrow range of flow conditions. Lack of information on model performance outside of these conditions limits the scenarios that can be reliably examined by the model.

Submersed Aquatic Vegetation

16. Some calibration of the SAV model appears to have occurred, but the details are not provided in the interim report. More detail will be necessary in the final report.

Creating a framework for model documentation that covers goals, assumptions, justifications for parameterization, calibration, and verification for this (and future) versions of the SAV model is good practice and will aid in the longevity and application of the SAV model. Based on Table 13 in BIO-WEST (2015), several parameters have been calibrated for two species that have been the focus of initial SAV model efforts. However, descriptions of the calibration approach and results have not been provided. The model development team is strongly urged to provide detail regarding their calibration plans.

17. Calibration and validation should consider efforts to compare model output of rates as well as state variables.

It is a common practice to use state variables such as biomass to validate numerical models of primary producers. However, especially in this SAV model where three critical rate processes are simulated (photosynthesis, respiration, growth), it is important to look at output from the model of these rate processes and to compare, in some fashion, these simulated rates to measured rates. If measured rates are not available for all species, their acquisition can be identified as a critical research activity to be done while the literature is scanned to provide some confidence regarding rates for some of the modeled species.

18. Developing an SAV model that can accurately simulate the observed maps of SAV coverage is unlikely and not advisable. Rather, validation exercises should be considered that take SAV coverage into account at larger spatial scales and compare patterns of SAV coverage between predicted and observed maps.

A "pattern-oriented approach" similar to that described by Grimm et al. (2005) could be considered for guiding model evaluation and validation. This approach also influences model development, but in a way that is complementary to the currently described efforts for the SAV model. This might include validation exercises comparing important patterns generated from the model that were not simply an outcome of the model inputs.

In addition to considering a pattern-oriented strategy, with a spatial scale of 1 m^2 and the decision to permit just one species per grid cell, it is highly unlikely that a simulated map of SAV coverage will directly mirror actual species distribution maps. A more realistic validation exercise might consider comparisons at the reach scale or some intermediate spatial scale above 1 m². Model evaluation should focus on aggregate measures (e.g., total biomass by type) and their seasonal and spatial patterns, rather than trying to match predicted and observed biomasses on a cell-by-cell basis within a survey. Validation should also consider the use of the SAV model as both a standalone model and in its role as generating habitat input for the FD model, to ensure that the appropriate aggregate measures are evaluated.

MODEL COUPLING

Model coupling is a special topic because of the plans by the model development team to use the results of the SAV modeling to provide habitat inputs to the FD model. Running models in a coupled mode involves additional issues beyond those identified above, which were based on running the two models independently (standalone).

There are four submodels within the overall ecological model: hydraulics (steady state 2-D model), water quality (QUAL2E), SAV, and FD. The hydraulics model is Dr. Thom Hardy's existing MD_SWMS model for both the Comal and San Marcos systems. The grid size is 0.25 m². Hydrology (flow and depth) is generated by having seven-day averages over the time period 2000 to 2013. The water quality model for both systems is QUAL2E. The model outputs from the hydraulics and water quality submodels that have been used to date in the FD model are maximum daily water temperature and minimum daily DO. The SAV model will require hydraulic and water quality model outputs of depth, temperature, and (eventually) some measure (e.g., average daily) of flow. There is currently no direct role for velocity or depth as inputs to the FD submodel.

The SAV submodel is a standalone model that can be used to examine questions related to HCP activities, and it is also planned to provide the habitat information for the FD model grid. While these are highly related uses of the SAV modeling, it is likely that compromises are needed in order for the same SAV model to be able to perform both uses. For example, trying to use the SAV modeling results as input to the FD model may push the SAV modeling to a finer spatial scale to match the FD model than if the only goal of the SAV modeling was to assess flow effects on SAV dynamics.

The actual coupling between the SAV and FD models is planned on being one-way, which is reasonable. That is, SAV affects FD, but FD does not affect SAV. This is biologically realistic and also allows for the SAV and FD models to be run separately if needed for computational reasons. The SAV modeling should generate outputs on spatial and temporal scales realistic for how FD uses these habitats within the model; that is, how do these habitats affect FD growth, mortality, reproduction, and movement on hourly to daily time steps for roughly 1-m^2 spatial resolution and within the FD model domain of a reach. For example, conversion of a grid cell to a different SAV species occurs just one time per month at this stage of model coupling. Careful consideration of whether this, as well as how other variables are transferred, is sufficiently accurate for use in the FD model is warranted. The SAV modeling is still unsettled, but it seems that a reasonable compromise can be found such that the SAV modeling can be used both for simulating SAV responses to flow and for providing habitat inputs to the FD modeling.

The use of steady-state hydraulics and dynamic QUAL2E as potential inputs to the SAV and FD model is reasonable provided the limitations of this coupling (hydraulics-QUAL2E) for use in the FD model are clearly detailed. The hydraulics model is used in two ways: a series of constant flows is simulated (steady-state for each flow) for direct use of depths and velocities in SAV and FD models, and using seven-day average values of flow (also to steady-state) as input to the QUAL2E model to generate hourly temperature and DO. The steady-state velocities and depths are re-gridded from the 0.25 m^2 of the hydraulics to the 1-m^2 grid of the FD model. The hourly temperature and DO are processed to obtain daily maximum temperature and daily minimum DO values. All of the FD model cells fall within a single QUAL2E segment, and thus the temperature and DO values in the appropriate QUAL2E segment are applied to all of the cells in the FD model. While the idea of model coupling is sound and seems simple and intuitive, the details are very important for conveying the limitations (and strengths like higher confidence) when the fully coupled set of models are used to simulate SAV and FD responses to HCP actions.

At every step of passing output from one model to be input to the next model in the chain, some information is lost (often variance) and the receiving model inherits the assumptions used to run the donor model. In the situation here, these steps include aggregation of 0.25 m^2 scale in the hydraulics to 1-m^2 resolution in the SAV and FD models, steady-state hydraulics used dynamically in the SAV and FD models, steady-state hydraulics used differently to generate velocities and depths versus as input to QUAL2E to generate temperature and DO, and all of the FD and SAV model cells being within a QUAL2E segment (i.e., no spatial variability). Careful evaluation and bookkeeping of the assumptions, of how information is generated (e.g., steady-state versus dynamic) and then passed to the next model, is needed to ensure the information from different sources is consistent and to know what types of scenarios can be realistically examined. The FD model will inherit the assumptions and limitations of all of the previous model analyses that provided inputs. Calibration and validation of standalone models independently do not guarantee they will perform with sufficient accuracy and precision when they are coupled.

CONCLUDING REMARKS

The ecological modeling is on a good pathway forward. The FD modeling has made significant progress toward the goal of predicting the effects of HCP actions on FD population dynamics. The SAV modeling is in an earlier stage of development and therefore its status is more difficult to assess. This review examined the available information and offers a suite of comments, some of which are conclusions and some of which are recommendations. The summary below is intended to help the modeling process continue toward its eventual objective of being a quantitative tool to assist in evaluating HCP-related actions on FD and SAV dynamics.

- Ensure adequate expertise is available to modify, run, and properly interpret the models once they are completed by the development team. [Comment 1]
- Utilize the power of the mechanistic approach embodied within the FD model by including the reasons that predicted responses occur; use the model in the inverse mode to identify key life stages, processes, locations, and timings for effective management actions. [Comment 2]
- Expand the factors explicitly included in the SAV modeling to include flow, and consider alternative formulations for dispersal and cell-level changes in SAV species that do not simply mimic the observed data but that depend on flow and other factors. [Comments 3 and 13]
- Keep track of the variance properties as information is passed from one model to the next. [Comment 4]
- Confirm the representativeness of the reach approach for FD so that results can be interpreted at the true population level that spans multiple reaches. [Comment 4]
- Plan for how to ensure that the SAV maps used in the FD model (either from the SAV model or uncoupled) can be used to predict habitat changes in response to flow. [Comment 5]
- Evaluate whether the growth, mortality, reproduction, and movement processes represented in the FD model should be (1) more linked to each other, which might lead to density-dependent responses; and (2) more linked to environmental variables such as flow. Logic charts showing how HCP actions can cause responses in the information passed from the hydraulics and water quality models to the SAV model; from the hydraulics, water quality, and SAV models to the FD model; and within the FD model itself, would benefit model communication and interpretation of the FD modeling results. [Comments 6, 8, and 9]

- Careful use of threshold-like formulations for processes in both models is needed because using minimum or maximum values of environmental conditions or cutoff values for SAV and FD variables can dampen responses to flow changes and generate sudden changes in SAV and FD model predictions. [Comments 7 and 13]
- Evaluate further the present assumptions about no nutrient limitation, the present formulation for light and temperature effects, and direct and indirect roles of flow in the SAV model. [Comment 10]
- Further confirm the calibration and the realism of the resulting parameter values, and the appropriateness of using a crop model for SAV using literature and site-specific information. [Comment 12]
- Ensure sufficient documentation/explanation of the SAV model and of the coupled version of the FD-SAV modeling for future evaluation and use of the models. [Comment 11]
- Expand on the calibration and validation of the FD model to address the partial tautological aspect of specifying the maximum densities from observed values and then showing the model replays total abundances, and the relatively low interannual variation of environmental conditions within the calibration time period. [Comments 14 and 15]
- Develop and implement a calibration and validation plan for the SAV model that includes model-data comparisons of biological rates and testing of the model's ability to produce key spatial patterns. [Comments 16, 17, and 18]

Much progress has been made, and there is still significant effort remaining in order to get the models to the point in their development and evaluation that they are ready for predicting responses of SAV and FD to HCP actions. These comments hopefully provide guidance for continuing on the path forward.

REFERENCES

BIO-WEST. 2015. Predictive ecological model for the Comal and San Marcos ecosystems project. Edwards Aquifer Habitat Conservation Plan. Interim Report. Contract No. 13-637-HCP.

EARIP (Edwards Aquifer Recovery Implementation Program). 2012. Habitat Conservation Plan. Edwards Aquifer Recovery Implementation Program.

Grimm, V., E. Revilla, U. Berger, F. Jeltsch, W. M. Mooij, S. F. Railsback, H.-H. Thulke, J. Weiner, T. Wiegand, and D. L. DeAngelis. 2005. Pattern-oriented modeling of agent-based complex systems: Lessons from ecology. Science 310(5750):987-991.

NRC (National Research Council). 2015. Review of the Edwards Aquifer Habitat Conservation Plan: Report 1. Washington, DC: The National Academies Press.

Parrish, R. H., and A. D. MacCall. 1978. The climatic variation and exploitation in the Pacific mackerel fishery (Vol. 167). State of California, Resources Agency, Department of Fish and Game.

Rose, K. A., J. H. Cowan, K. O. Winemiller, R. A. Myers, and R. Hilborn. 2001. Compensatory density dependence in fish populations: Importance, controversy, understanding and prognosis. Fish and Fisheries 2:293-327.

Teh, C. 2006. Introduction to Mathematical Modeling of Crop Growth. Boca Raton, FL: Brown Walker Press.

Appendix B

Biographical Sketches of Committee Members and Staff

Danny D. Reible (NAE) is currently the Donovan Maddox Distinguished Engineering Chair at Texas Tech University. He previously served as Director of the multi-university consortium, the Hazardous Substance Research Center South and Southwest (1995-2007), while at Louisiana State University and as the Bettie Margaret Smith Chair of Environmental Health Engineering (2004-2013) and director of the Center for Research in Water Resources (2011-2013) at the University of Texas. Dr. Reible was inducted into the National Academy of Engineering in 2005 for his work in identifying management approaches for contaminated sediments. He has led the development of in-situ sediment capping and has evaluated its applicability to a wide range of contaminants and settings, including polycyclic aromatic hydrocarbons from fuels, manufactured gas plants and creosote manufacturing facilities, polychlorinated biphenyls, and metals. His current research activities are focused on sustainable water management and the assessment and remediation of contaminated sites. He is a Fellow of the American Institute of Chemical Engineers and the American Association for the Advancement of Science. He received his B.S. from Lamar University, and his M.S. and Ph.D. in chemical engineering from the California Institute of Technology.

Jonathan D. Arthur is the State Geologist of Florida and Director of the Florida Geological Survey, a division of the Florida Department of Environmental Protection. Dr. Arthur received his B.S. and Ph.D. from Florida State University and is a Fellow of the Geological Society of America. He has served as past president of the Association of American State Geologists and the Florida Association of Professional Geologists, and presently serves on the Florida Board of Professional Geologists. He also served on numerous committees related to restoration of the Florida Everglades. His research has involved aspects of hydrogeology and hydrogeochemistry, including hydrogeologic framework mapping, aquifer vulnerability modeling, and aquifer storage and recovery, the latter with emphasis on water-rock interactions and fate of metals and metalloids during variable oxidation-reduction conditions. Dr. Arthur was a member of the Academies' Committee on Sustainable Underground Storage of Recoverable Water.

M. Eric Benbow is an associate professor of entomology at Michigan State University. His research involves basic and applied multiple-scale studies on the biology and ecology of aquatic ecosystems, how terrestrial and aquatic ecosystems are coupled, the influence of human activities on those processes, and microbe-insect interactions in aquatic systems and carrion decomposition. Specific projects include the ecology of microbial-invertebrate interactions and their role in mycobacterial disease emergence in West Africa; microbial-insect carrion interaction networks in watersheds of southeast Alaska; watershed biomonitoring; and carrion decomposition with applications in forensics, including human postmortem microbiome studies. He has studied water withdrawal and watershed development in the tropics, including monitoring how invertebrate communities respond to these impacts. Dr. Benbow has served as a consultant to the World Health Organization on Buruli ulcer, the Republic of Palau for stream bioassessment, and the New Jersey Forensic Science Commission, Forensic Anthropology and Associated Forensic Specialties Sub-Committee; as an expert witness in a contested case involving Hawaiian streams; and as an Executive Committee member and former president of the North American Forensic Entomology Association. He received his B.S. and Ph.D. in biology from the University of Dayton.

Robin K. Craig is the William H. Leary Professor of Law at the University of Utah College of Law. Her research focuses on "all things water," especially the impact of climate change on freshwater resources and the oceans and the intersection of water and energy law. She has just published a water law textbook, *Modern Water Law: Private Property, Public Rights, and Environmental Protections,* and authored the chapter on the Endangered Species Act and the chapter on constitutional takings, both of which prominently featured cases and commentary from the Edwards Aquifer. Dr. Craig previously taught at the Lewis & Clark School of Law; Western New England College School of Law in Springfield, Massachusetts; Indiana University-Indianapolis School of Law; and the Florida State University College of Law in Tallahassee, Florida. She served on three successive Academies' committees on the Clean Water Act and the Mississippi River. She is also active in the American Bar Association's Section on Environment, Energy, and Resources, where she just completed a three-year term on the Executive Council and where she currently serves as Co-Chair of the Water Resource Committee. She received her BA from Pomona College, her M.A. from Johns Hopkins University, her Ph.D. in English literature from UC Santa Barbara, and her J.D. from Lewis and Clark College.

K. David Hambright is a professor of biology and Director of Environmental Studies at the University of Oklahoma. During the past decade his research has centered on the ecology, evolution, and management of the invasive and toxigenic golden alga, *Prymnesium parvum*, in lakes and rivers in Oklahoma, Texas, West Virginia, and Pennsylvania. He has recently begun a new long-term research effort aimed at coupling satellite-based remote sensing, digital field photography, and long-term water quality monitoring data on Oklahoman lakes in the effort to develop real-time monitoring capabilities aimed at ensuring public safety on the many public-access recreational lakes in the state. His expertise includes research in climate change and water quality interactions, wetland restoration and habitat and species conservation, paleolimnology, ecosystem modeling, and biodiversity, as well as experience in working with diverse research and modeling teams, interest groups, and stakeholders in politically sensitive systems. He received his B.S. in biology from the University of North Carolina-Charlotte, his M.S. in biology from Texas Christian University, and his Ph.D. in ecology and evolutionary biology from Cornell University.

Lora A. Harris is an associate professor at the University of Maryland Center for Environmental Science, based at the Chesapeake Biological Laboratory. She is an estuarine ecologist who applies field and modeling approaches to address important questions regarding nutrient dynamics, primary production, and ecosystem structure and function in a range of estuarine ecosystems. She is interested in climate impacts on estuaries and lagoons, with a particular focus in salt marsh and sea grass ecosystems. Some of her most recent work has involved participatory modeling efforts involving stakeholders and managers seeking solutions to improve water quality and restore seagrasses in Delmarva coastal lagoons and a collaboration with wastewater engineers to understand the restoration trajectories of hypoxic estuaries. Dr. Harris works closely with state and regional agencies in both a research and an advisory capacity. She received her B.S. from Smith College and her Ph.D. from the University of Rhode Island.

Timothy K. Kratz is the director of Trout Lake Station and the Center for Limnology at the University of Wisconsin-Madison. He is currently on a rotation at the National Science Foundation where he is serving as a program officer in the NEON Science and Macrosystems Biology program in the Biological Sciences Directorate. His research interests include the long-term, regional ecology of lakes; metabolism and carbon dynamics of lakes; land-groundwater-surface water interactions; global patterns in lake dynamics through development of the Global Lake Ecological Observatory Network. He has served on four Academies' committees, including the Committee on Grand Canyon Monitoring and Research. He earned his B.S. in botany from the University of Wisconsin, his M.S in ecology and behavioral biology from the University of Minnesota, and his Ph.D. in botany from the University of Wisconsin.

Andrew J. Long is a research hydrologist and the groundwater specialist for the U.S. Geological Survey, Washington Water Science Center. His work has involved development and computer coding of mathematical models to understand and quantify dual-porosity flow and transport of solutes and heat in karst aquifers, such as the Edwards. Inverse modeling and uncertainty assessment in modeling has been an important component of his research program. His current research involves groundwater age dating, lumped-parameter models, aquifer classification, hydrochemical evaluation, heat transport, and groundwater recharge, using such methods as dye tracing, age-dating tracers, geophysical methods, and hydraulic aquifer testing. Prior to coming to the USGS, Dr. Long worked for the South Dakota Water Department to simulate groundwater flow in the karstic Madison aquifer using MODFLOW, and he worked as a consultant conducting GIS analysis and groundwater modeling with MODFLOW related to a proposed low-level nuclear waste site in Boyd County, Nebraska, and to assess ammonia contamination for Terra Nitrogen in Sergeant Bluff, Iowa. He recently served as an adjunct assistant professor at the South Dakota School of Mines and Technology, where he received his B.S., M.S., and Ph.D., all in geological engineering.

Jayantha Obeysekera is the chief modeler at the South Florida Water Management District, where he established and managed a group of about 60 modelers covering hydrologic, hydrodynamic, water quality, and ecological disciplines. He has more than 25 years of experience practicing water resources engineering with emphasis on both surface water and groundwater modeling, and implications of climate variability in planning and operation of complex water resources systems. He was a co-principal investigator for a National Science Foundation–funded project on the investigation of the tsunami impacts on coastal water resources in Sri Lanka. Dr. Obeysekera also served as an external agency member to the U.S. Army Corps of Engineers to review post-Katrina hydrologic modeling of the greater New Orleans metropolitan area. He has served on three Academies' committees, including the Committee on Sustainable Water and Environmental Management in the California Bay-Delta, which had a Habitat Conservation Plan as its central focus. Presently, he is serving as a member of the National Climate Assessment and Development Advisory Committee. Dr. Obeysekera holds a B.S. in civil engineering from the University of Sri Lanka, M. Eng. from University of Roorkee, India, and a Ph.D. in civil engineering from Colorado State University. He is a registered professional engineer in the state of Florida and has been appointed as an affiliate research professor at the Florida Atlantic University.

Kenneth A. Rose is the E. L. Abraham Distinguished Professor in Louisiana environmental studies at Louisiana State University. His current research is focused on modeling population dynamics of fish and aquatic food webs, and how they respond to a variety of types of stressors, including changes in water flows and quality, lethal and sublethal effects of contaminants, hypoxia, alteration of physical habitat, and climate change. He recently published a model of the population dynamics of the delta smelt, which is a listed species in the California Delta that is a center of controversy about how much water can be pumped out of the system for irrigation and water supply, and he has also published on lower trophic level (algae and micro and macro zooplankton) food web dynamics. Dr. Rose was a member of review teams for several biological opinions involving delta smelt and salmon. He has served on two Academies' committees, including the Committee on Sustainable Water and Environmental Management in the California Bay-Delta that evaluated the mitigation and conservation actions of biological opinions and the science underlying the short-term and long-term environmental and water usage decision-making of the system. He received his B.S. from SUNY Albany and his M.S and Ph.D. in fisheries science from the University of Washington.

Laura Toran is the Weeks Chair in Environmental Geology at Temple University in Philadelphia. She has 30 years of experience in modeling and monitoring groundwater. Her recent research activities include using karst springs to understand transport in karst, monitoring urban stormwater and streams, and developing hydrogeophysical techniques to predict groundwater-surface water interaction. She teaches classes in groundwater hydrology including modeling with MODFLOW. She served on the Academies' Committee on Opportunities for Accelerating Characterization and Treatment of Waste at DOE Nuclear Weapons Sites. Dr. Toran received her B.A. in geology from Macalester College and her Ph.D. in geology from the University of Wisconsin.

Greg D. Woodside is the executive director of Planning & Natural Resources at Orange County Water District. Mr. Woodside has 25 years of experience in water resources management and hydrogeology. Mr. Woodside is a registered geologist and certified hydrogeologist in California. Mr. Woodside oversees the Planning and Watershed Management Department and the Natural Resources Department at the Orange County Water District. Staff in these departments prepare the District's environmental documents, permit applications, Groundwater Management Plan, and Long-Term Facilities Plan, and conduct the District's natural resource management, watershed planning, and recharge planning. In particular, he has evaluated conjunctive use and Aquifer Storage and Recovery projects in the Orange County Groundwater Basin, Central Basin, and San Gabriel Basins, including projects that would recharge up to 50,000 acre-feet per year of recycled and imported water. Methods used by Mr. Woodside to evaluate conjunctive use projects include integrated surface and groundwater budgets, flow path analysis, simple analytical models of groundwater flow, and complex three-dimensional numerical models. He holds a B.S. in geological sciences from California State University, Fullerton, and an M.S. in hydrology from the New Mexico Institute of Mining and Technology.

STAFF

Laura J. Ehlers is a senior staff officer for the Water Science and Technology Board of the National Academies of Sciences, Engineering, and Medicine. Since joining the Academies in 1997, she has served as the study director for more than 20 committees, including the Committee to Review the New York City Watershed Management Strategy, the Committee on Bioavailability of Contaminants in Soils and Sediment, the Committee on Assessment of Water Resources Research, the Committee on Reducing Stormwater Discharge Contributions to Water Pollution, and the Committee to Review EPA's Economic Analysis of Final Water Quality Standards for Nutrients for Lakes and Flowing Waters in Florida. Ehlers has periodically consulted for EPA's Office of Research Development regarding their water quality research programs. She received her B.S. from the California Institute of Technology, majoring in biology and engineering and applied science. She earned both an M.S.E. and a Ph.D. in environmental engineering at the Johns Hopkins University.